SYNERGETICS:

An Adventure in Human Development

SYNERGETICS:

An Adventure in Human Development

N. Arthur Coulter, Jr., M.D.

Prentice-Hall, Inc.   Englewood Cliffs, N.J.

Prentice-Hall International, Inc., *London*
Prentice-Hall of Australia, Pty. Ltd., *Sydney*
Prentice-Hall of Canada, Ltd., *Toronto*
Prentice-Hall of India Private Ltd., *New Delhi*
Prentice-Hall of Japan, Inc., *Tokyo*

©1976 by

Prentice-Hall, Inc.
Englewood Cliffs, N.J.

Library of Congress Cataloging in Publication Data

Coulter, Norman Arthur
   Synergetics : an adventure in human development.

   Includes index.
   1. Self-actualization (Psychology) 2. System
theory. I. Title. [DNLM: 1. Mental processes.
BF455 C855s]
BF637.S4C68   158   75-39365
ISBN 0-13-879981-4

Printed in the United States of America

To Betty and Bob

PREFACE

This book is the result of nearly thirty years of search and research. The journey began August 6, 1945 with the atomic bombing of Hiroshima. Like many Americans, I was appalled. It seemed to me then, and still does today, that the bomb should not have been treated as "just another weapon"; that, at the very least, a demonstration could and should have been arranged for Japanese leaders at some uninhabited area. Such a demonstration would have been just as convincing as the actual destruction of a city. The war would have ended just as soon, and America's honor would not now have this ignoble stain. As it is, a war that began in one day of infamy ended in another.

But never mind that. Far more important was the realization of the terrible significance of nuclear weapons for the future of humankind. It was obvious that a nuclear arms race would soon follow, leading inevitably to the insane balance of terror that now exists, ever more dangerous in the face of the smug complacency of our leaders. Sooner or later, some idiot will push the button.

But this is not another book about the danger of the nuclear sword of Damocles. I cite these facts merely to explain the motivation for my long search. It then seemed clear that the only real solution to this problem was the abolition of war itself. Slavery and cannibalism have been eliminated—why not war?

It was evident that the United Nations could not prevent war—and indeed, it has not. This is not said to belittle the United Nations, which does many good things that are too little known. But the imperatives of fear, hate, greed and the lust for power that drive men and nations to war are too strong. Something much more basic is required.

7

I concluded that *what is needed is nothing less than a major evolution of the human mind, which would give the rational, humane part of the mind a much greater control over the emotional part.* Such a cultural mutation could, I felt, be brought about. Today I am more than ever convinced that it can. Of course, it won't be easy!

And so began my long search. I do not claim to have found the answer. Perhaps this book will contribute to the solution. This is for you, the reader, to judge.

The need for a cultural mutation of the human mind is not diminished by the fact that several new crises have emerged today that I did not then foresee. All my generation had to worry about was the threat of nuclear holocaust; the young today also face the specters of mass starvation, runaway population growth, the not-distant prospect of exhaustion of nonrenewable resources, the still uncontrolled and increasing pollution of air, water and oceans. This multiple crisis makes such a cultural mutation of the mind even more urgent. For the problem is not that rational men and women do not clearly see what needs to be done. The problem is that irrational social forces, political institutions, economic drives and cultural obstacles are too strong; they completely block rational efforts to resolve the crises.

It has indeed been a long journey with many false trails, egregious blunders and agonizing failures. But it has also been an exciting journey, with enough "bursts of insight" and "breakthroughs" to have made the adventure worthwhile. I am far from satisfied with the result. But I hope that others will correct the flaws I have overlooked.

N. Arthur Coulter, Jr., M.D.

Acknowledgments

Seldom has one book owed so much to so many. The scholar will quickly note the debt to such fields as psychoanalysis, analytical psychology, Gestalt psychology, cybernetics, general semantics, dianetics, Buddhism, Christianity, Western philosophy, Eastern mysticism, sociology, computer science and many many others, both orthodox and heterodox. My impatience to present these ideas and techniques in a form unencumbered by a mass of scholarly references has caused me to slight the contributions of many. For this, I humbly apologize.

The journey has also been made much easier by the help, criticism, evaluation, testing and other assistance of many friends who joined me in the adventure. My deepest thanks go to Betty Ash, James Baker, Dr. E.R. Barenburg, Dr. D.W. Batteau, Donald and Ann Benson, Waldo Boyd, Pat Buchan, John Carey, Professor Ed Cole, Paul and Mary Cox, James and Marguerite Craig, Lorraine Cullen, Bob Gardner, Dr. Joseph K. Geiger, William Hamilton, Charles Hampden-Turner, Lee Henrichs, Alma Hill, Fred and Barbara Hibbard, Bob and Betty Howard, Rev. William Jefferys, Frida Kaech, Ben Keller, Henry Lahann, Sadah Loomis, Margaret Mason, Professor W.S. McCulley, Richard McMahan, Margaret Meade, Edward J. Michaels, Reynolds Moody, Steve Moore, Richard L. Morris, Lew Mortensen, Robert Nichols, James and Frances Norris, Bill Perlman, Don Purcell, A.W. Schrader, Jr., Bill Sell, Nancy and Walter Strode, J.H. Sydell, George Tullis, Helen Urquhart, Bob Van Nice, A.E. Van Vogt, Florence Worrell, Orpha York, and James Young.

My special thanks go to Robin Mays, who patiently deciphered my scrawl to type the manuscript, and to Gwynn Moore who did the illustrations.

Last and most, my special appreciation goes to my wife, whose tolerance and understanding at critical moments have meant so much.

CONTENTS

The idea I find in synergy: it should be as simple for a child to go from New York to Los Angeles as loved and well-cared for as he would be in going from room to room in his home... my way home goes by your house... hands and I are music brothers.

——Wayne Batteau

CHAPTER 1

WHAT IS SYNERGETICS?

There is available to every human mind a state of advanced consciousness and well-being that is exciting, vigorous and incredibly beautiful. It is characterized by an expansion of awareness, by an enhancement of rationality and by a remarkable phenomenon called think-feel synergy. This state is called the *synergic mode of function.*

The word "synergy" means, literally, "working together." In medicine, it has long been used to denote the working together of two or more drugs, or of two or more muscles acting about a joint. Applied to the human mind, "synergy" denotes the working together of the enormous variety of functions that comprise the mind, producing a new whole that is greater than the mere sum of its parts.

When the synergic mode turns on, the mind lights up. Perceptions grow more vivid and acute, with "flash-grasp" of complex situations a not infrequent occurrence. Thinking becomes faster, more accurate and remarkably *clear.* Often thought-trains race along several tracks at once. Actions become more apt and multipurposed, with a high gain-to-effort ratio. Emotional tone ranges from cheerfulness to enthusiasm, with a harmonious blending of thought and emotion that is highly exhilarating. Abilities long dormant or even unsuspected are activated as the wave of synergy surges into the hidden depths of the mind.

So much goes on so fast that it is impossible to describe it adequately; the concrete experience is so vivid and multifaceted that it evades all efforts to define it in words. But there is one remarkable feature of the synergic mode that stands out above all others and that bodes well for the future of humanity: *it is literally and logically impossible for a person operating in the synergic mode to perform an unethical act.*

The reason for this is easy to see. Synergy involves the working together of the parts of any complex system; and each person is not only an individual, but a part of the various groups and organizations to which he belongs, and to society as a whole. In the synergic mode, a human being acts naturally so as not only to achieve his own goals, but also, wherever feasible, to promote the goals of others, with least impedance to anyone. The Golden Rule—"Do unto others as you would have them do unto you"—becomes not a moral commandment to be obeyed, but a natural and logical consequence of his mode of being, as natural as breathing, sleeping or sexual activity.

The prevailing outlook of a synergic being may be described as one of *synergic altruism.* He is not selfish, as this is commonly understood, but always considerate of the needs and interests of others, and ever ready to engage in cooperative enterprise. On the other hand, he is not selfless, sacrificing himself needlessly for others; he selects his own goals and pursues them vigorously, overcoming obstacles in his way.

The synergic mode of function has been experienced by many people in the course of history; doubtless many readers have felt it turn on at one time or another, without identifying it as synergy. But most would agree that it is not characteristic of the average person today. When I first experienced the synergic mode, early in 1952, I wondered about this. And I was disappointed that the state did not last. This led to the concept of a science of synergy—namely, synergetics.

Synergetics

A science of synergy has tremendous implications for mankind. From the standpoint of the individual, it will enable him to turn on the synergic mode whenever he chooses. More than this, it will enable him to *stabilize* in the synergic mode, so that he can be this way almost continuously. For himself, this will naturally lead to greater effectiveness and well-being.

From the standpoint of society, the implications of synergetics are even greater. Imagine a society in which everyone followed the Golden Rule, not because he was "trying to be good," but because it was his nature to do so! The age-old dream of a true brotherhood of man would be a living reality.

But the implications of synergetics are even greater than this. We have thus far considered only the synergic mode as it manifests itself in individual human minds. But synergetics is potentially applicable to any complex system. In principle at least, it is possible for any such system to operate in the synergic mode.

This means, for example, that synergetics can be applied to small human groups. Even if some of the members of the group are not *individually* synergic, the group as a whole may operate in the synergic mode. Again, this sometimes happens naturally, for example in some athletic teams. But a science of synergy will enable any group that so chooses to achieve this state. This branch of synergetics may be called Group Synergetics.

Similarly, synergetics may be applied to larger groups, organizations, communities, even whole societies. Imagine a whole community operating in the synergic mode! The prospects are truly mind-boggling. Social Synergetics, although admittedly the most difficult, is probably the most important branch of synergetics.

Individual Synergetics, Group Synergetics and Social Synergetics, then, are three vitally important divisions of this field. Each can and necessarily will be developed separately. But the idea of synergy again applies. Each of these three branches can be applied in such a way that they mutually reinforce one another. Thus, for example, a group may more readily operate in the synergic mode if each member of the group is himself synergic. Conversely, being a member of a synergic group makes it easier for an individual to turn on the synergic mode. Similarly, synergic relations exist between Social Synergetics and Individual and Group Synergetics, respectively. The three together form a synergic whole.

Many other branches of synergetics are possible, of which one deserves special mention. It has often been noted that the increasing specialization in science (and other forms of human endeavor) creates problems of communication of growing severity. A need exists for *generalists*—people able to bridge the gap between fields. Synergetics, applied to the problem of integrating different fields of knowledge, may facilitate the work of generalists. We will not, however, consider this further here.

My original impetus for contributing to a science of synergy was the dropping of atomic bombs on Hiroshima and Nagasaki, which made it clear that war must be abolished. Some years later I was joined in this enterprise by a group of friends and colleagues—scientists, intellectuals, housewives, workers, persons from all walks of life. Together, we have tried to develop a science of synergy. A preliminary exposition of synergetics was published in 1955; but this book is a definitive presentation of what we and others have found thus far.

Techniques have been developed that enable a person of at least average intelligence and self-honesty to turn on the synergic mode in himself. These techniques are relatively easy to learn and to use, requiring an effort comparable to that of learning to drive a car or to play simple pieces on a piano. Drugs or hypnosis are not used; indeed, it is recommended that they not be combined with synergetics.

Techniques have also been developed for evoking the synergic mode in small human groups, and contributions have been made to the field of Social Synergetics.

It should be emphasized, however, that the importance of synergy has been recognized independently by other workers; indeed, in recent years there has been a burst of activity in this field. As already noted, the concept of synergy has long been used in medicine. An early pioneer in the application of the concept of synergy to sociology was the American sociologist Lester Ward. (1)* Ward used the term to denote the interaction of social forces that were in themselves destructive, but which through mutual checks and constraints had a constructive effect.

Another early pioneer in synergetics was Buckminster Fuller, whose concepts of anticipatory design for spaceship earth have inspired generations of students. Fuller (2) emphasized the key role of synergy in producing a whole that is greater than the sum of its parts. Something new emerges—something that can exist only at the higher level of organization made possible by synergy. A musical chord, for example, is more than just several notes sounded together; it is an emergent whole, having a quality that is absent from the notes sounded separately. The pieces of a jigsaw puzzle, randomly arranged, are simply a collection of material objects. When they *fit together,* a new whole manifests itself in the painting—a whole that did not exist until the puzzle was "solved." Similarly, when the synergic mode turns on in a human mind, a new, higher level of function comes into being. It is sometimes difficult to appreciate this later if the synergic mode turns off.

---

*A number in parentheses following material derived from or referring to another source indicates the source reference in the Notes and References section at the back of this book.

Another pioneer in synergetics was Ruth Benedict. In a series of unpublished lectures given at Bryn Mawr College in 1941, Benedict introduced the concepts of high synergy and low synergy. She defined these terms as follows: "I spoke of societies with high social synergy where their institutions insure mutual advantage from their undertakings, and societies with low social synergy where the advantage of one individual becomes a victory over another, and the majority who are not victorious must shift as they can."(3)

Benedict's ideas were later rescued by Abraham Maslow, who preserved the only manuscript of her work. Maslow, one of the founders of humanistic psychology, did much to bring the concept of synergy to the attention of the literate public.(4) His many other contributions are too well-known to need recounting here; but it is worth noting that such Maslowian concepts as that of "self-actualization" and the "peak experience" aptly characterize major facets of what we here call the synergic mode.

Major progress in Social Synergetics has been made recently by several workers and groups. James and Marguerite Craig, social psychologists, have introduced the concept of *synergic power* as an alternative to coercive power and an antidote to the sense of powerlessness so many feel today.(5) They show that power in itself is neither good nor bad; it depends on how it is used and what it is used for. More important, they show how it is possible for people who lack coercive power or who dislike using it to create, develop and apply synergic power. They define synergic power as "the capacity of an individual or group to increase the satisfactions of all participants by intentionally generating increased energy and creativity, all of which is used to co-create a major rewarding present and future." Some of their techniques are described later in this book.

Charles Hampden-Turner has applied the idea of synergy to the development of a strategy for poor Americans, which would enable them to bootstrap themselves out of their predicaments.(6) Emphasizing that synergy is a complex idea with many facets, he selects three interdependent definitions that are mutually synergic:

(a) Synergy is the fusion between different aims and resources to create MORE between the interacting parties than they had prior to the interaction.

(b) Synergy is created by the resolution of apparent opposites and social contradictions.

(c) Synergy grows out of a dialectical and dialogical process of balance, justice, and equality, between persons or groups, and between the ideas and resources they represent; such synergy always exists on multiple levels.(6)

He points out that the many facets of synergy are an advantage in a social situation, because each person or group can find at least one of these facets to relate to.

Among groups that emphasize synergy, there is first of all the Hawaiian Health Net, monitored by Walter and Nancy Strode. Formed in 1972, the net has grown rapidly, involving mainlanders as far away as the East Coast. Inspired by "the image of man-come-of-age—man with knowledge of how his world works, man aware of his interdependence with the rest of life, man with a vision of becoming fully actualized, at-one, healthy, whole," this evolving network has been trying to develop a new, positive concept of health—health as something more than the mere absence of disease. In conferences, workshops, and other modes of communication they are searching for synergies among various viewpoints as a means to this end.

Another group is the Committee for the Future, led by Barbara Marx Hubbard and John Whiteside. This group sponsors a new type of conference, called a SYNCON (standing for *Syn*ergistic *Con*-vergence). The basic idea is to eliminate the adversary mode of conflict resolution and replace it by a holistic approach that takes into due account the diverse interests and, as all participate in the decision-making process, creates a focus of hope. To facilitate the process, a large wheel is constructed, divided into segments within which participants having particular interests and skills may interact. Later, "the walls come down." The first SYNCON was held at Southern Illinois University in 1972. Since then a growing number have been held in other cities.

Mention should also be made of the work of Donald Benson, Assistant Professor of Synergetics at Shaw University, on a synergetic approach to a learning society; Family Synergy, a California group pioneering in the development of synergic lifestyles, based on the idea of community rather than the "nuclear family"; and SYNERGY ACCESS, a newsletter/service monitored by Wes Thomas and devoted to facilitating communication and contact among people interested in synergy. The early 1970s have been marked by the independent discovery of the tremendous potential of synergy by a growing number of individuals and groups. If this trend continues, the 1970s may come to be known as the Synergic Decade, the beginning of a new Age of Synergy for humankind.

Indeed, synergy is of universal applicability. It can be applied not only to the human mind, but also to human groups, to organizations of all kinds, to industrial enterprises, to entire economic systems, to international relations, indeed, to any functioning system. Its basic characteristics of improving effectiveness by mutual aid, and of producing emergent wholes, can be brought about in every system to which it is applied.

From the standpoint of the original motivation of my research, it seems clear that think-feel synergy in the minds of the leaders of nations, and synergy among the interactions between nations, would lead to the abolition of war and the destruction of all nuclear weapons.

Dysergy

But the idea of synergy, itself, clearly is not enough. The leaders of nations pay lip service to international cooperation, but they still wage war. And in human organizations, groups, and the minds of individuals, synergy is a comparatively rare event.

What's the problem? Why is synergy so difficult to achieve, and to maintain?

In basic terms, the reason for this is the existence of interactions that are the very opposite of synergy—interactions that promote one function but impede another. The system works against itself, tying up a great deal of energy with much spinning of wheels, grinding of gears, and gnashing of teeth. We refer to such a phenomenon as *dysergy.*

The term "dysergy" is a coined word meaning "difficult working." It is used in place of "conflict" because conflict is not necessarily dysergic. The conflict between athletes, for example, can at least sometimes challenge the competing athletes to peak performance. At other times, such conflict may be dysergic.

Dysergy is to be found everywhere in human activities. Few individuals are free of dysergy in their own minds. Some dysergy is present in even the most synergic human groups; and there is a lot of dysergy in almost every group. The organizations, institutions, economic systems, political entities, and other functional structures of mankind are all beset by dysergy in a variety of forms.

To achieve and maintain synergy, it is clearly necessary to eliminate dysergy—or at least to reduce it to manageable proportions. Accordingly, a considerable amount of research has been devoted to the development of ideas and techniques for the elimination of dysergy.

Synergetics may be defined as the art and science of producing synergy and reducing dysergy in complex systems. Applied to the minds of individuals, it enables people to live happier, more effective lives. Applied to groups, it promotes the development of a high degree of mutual understanding, trust, teamwork, and love. And while its application to larger institutions has been minimal to date, there is every reason to believe that highly rewarding results may be achieved.

Indeed, the combined process of producing synergy and reducing dysergy is itself synergic. For the more synergy is produced, the easier it is to reduce dysergy. Conversely, the more dysergy is reduced, the easier it is to produce synergy. A cycle can be set up which, if maintained, results in synergic growth and development.

What This Book Is About

This book is divided into four parts. The first part deals with Basic Synergetics—ideas and tools that are potentially applicable to any functioning system. Basic Synergetics serves both as an introduction to other parts of the book and as a basis for developing applications to special fields.

The second part deals with Individual Synergetics. It contains ideas and techniques for promoting synergy and eliminating dysergy in the minds of individuals. By applying these ideas and techniques—either alone or with the help of a friend of like interests—the individual can solve personal problems, eliminate emotional hangups, and amplify his intelligence and effectiveness. More than this, synergetics turns on the synergic mode. A procedure enabling an individual of at least average intelligence to stabilize in the synergic mode is included as an Appendix.

The third part deals with Group Synergetics. This consists of ideas and techniques for promoting synergy and reducing dysergy in small human groups. The synergic potential available to a group that is organized and functioning according to synergetic principles is enormous. This is especially so when the members of the group are also aware of and practice Individual Synergetics. The principle of synergy again applies: participation in a synergetic group helps the individual in his own synergic evolution, and the practice of Individual Synergetics by members of the group enhances group synergy.

Fourth, there is Social Synergetics. This is of particular interest to individuals and groups struggling to achieve freedom from and control over the huge institutions that everywhere dominate and oppress us.

Finally, it should be emphasized that synergetics belongs to the people, not to any profession or organized group. The ideas and techniques here described are easy to learn and to use, and require no special knowledge beyond that contained in this book and in the mind of the average reader. No special apparatus is required (a tape recorder is sometimes useful but not essential), and drugs and hypnosis are *not* used. Those of us who have participated in the development of these techniques are especially concerned that they not be exploited commercially. Freely helping one another help himself—that's synergetics in action.

PART ONE

Basic
Synergetics

CHAPTER 2

THE MODE LADDER

In this chapter, we present one of the basic ideas of synergetics—the Mode Ladder.*

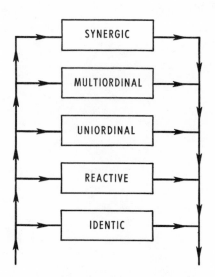

THE MODE LADDER

It will be convenient to describe this idea in terms of the human mind, to which it is directly applicable. However, it also may be applied to human groups, to organizations, to institutions, to economic systems, to political systems, to ecosystems, and so on.

Each rung of the Mode Ladder is a *mode of function* available to the human mind. The inputs to the ladder are the sense inputs—vision, hearing, smell, taste, touch, skin temperature, internal body sensations, etc. The outputs are the actions and communications of which a human being is capable.

At any given moment, an individual usually has a *prevailing* mode

---

*This idea was inspired by Hughling Jackson's concept of "levels of function" in the human nervous system.

of function—one that characterizes the way he thinks and functions mentally. Actually, this is oversimplified; most or all modes are simultaneously active to a degree. But one is usually dominant.

Just as an emotional mood can change from time to time, so can a person's mode change. On some occasions he may be near the top rung, perhaps even synergic. On others, he may be reactive or lower. Mode fluctuates from hour to hour, day to day, week to week, situation to situation. But at any given moment, it is usually possible to judge what a person's mode is.

Let us now consider each rung of the ladder in turn, from the bottom up.

Identic Mode. This is the lowest rung of the ladder. This is the mode of a person who is under hypnosis. It is also the mode of a mind in magical thinking. The conditioning process that occurs when a Pavlovian conditioned reflex is established is part of this mode.

The common denominator of these states is a mental operation called *identification*. One mental content is linked to another, as if a switch were turned on connecting the two. Thereafter, whenever the first content occurs, the second follows and *is equated to the first*. There is no discrimination. There is no awareness as we understand it. If "peaches" is linked to "cream," in the Identic Mode "peaches" = "cream." They are treated as being *identically the same*.

Thus, for example, the hypnotist says, "At the count of three you will open your eyes and see nothing." He counts, and the subject opens his eyes and literally sees nothing. The words of the hypnotist have been linked to a shut-off of visual perception.

The witch doctor makes a voodoo doll of his enemy M'Komba. He sticks a pin in the chest of the doll and M'Komba feels a sharp pain in his chest. This is magical thinking—identification without discrimination or awareness.

The TV announcer says, in a loud and confident manner, "Blubberin works faster than aspirin." He says it over and over, perhaps with a model of two stomachs showing Blubberin dissolving faster. Later, a viewer sees Blubberin in a store and buys it because it works faster than aspirin.

A bell is rung. Meat is placed in a dog's mouth. Saliva flows. This is repeated several times. Then a bell is rung, but no meat is given. After a precise interval, saliva flows. A conditioned reflex has been established—a *linkage* between the sound of the bell and the flow of saliva.

And there is the man who, whenever a water faucet is turned on, finds, a short time later, that he has to go to the bathroom.

The Identic Mode is present in everyone. Anyone can be hypnotized if he agrees. But even when awake, all of us are somewhat responsive to suggestion, verbal or nonverbal. Advertising *works*. So does propaganda. Most of us would deny that *we* are identic, without awareness, without discrimination. But a close examination will show the existence and occurrence of identifications. These are especially evident when a person does the same thing day after day.

Reactive Mode. This is the mode of a person who is *driven* by emotion—fear, anger, hate, grief, greed, lust, guilt. The individual is aware and he does make simple distinctions. These are characteristically *diametrically valued*—good or bad, right or wrong, black or white. They are *rigidly* held. Evidence to the contrary is explained away, rejected, or ignored. There is no discrimination of degrees between extremes.

Emotional reactions can occur in everyone. When they do, reactive thinking is easy to observe. What is mine is good, pure, right. What is against me is bad, soiled, evil. Sometimes reactive thinking is cast in logical form. Reason is "good," and therefore the forms of reason are used. But they are used to prove or demonstrate a belief that is already accepted, and that will be rigidly held, no matter what! ("My mind is made up; don't confuse me with facts.") It is a sad commentary on the human condition, but much of human thinking is reactive in mode. This is especially true in politics!

It should be emphasized that we do not consider all emotions as "bad." Emotions are part of human nature; they give experience its meaning and life its zest. It is quite possible for a person to be emotional and not reactive in mode.

Uniordinal Mode. The next rung on the Mode Ladder is the Uniordinal Mode. In this mode of function, the mind not only distinguishes between extremes, but also differentiates degrees between extremes. Ideas are not so rigidly held, and the mind can change if an inconsistency is shown or a contradictory fact is presented. The mind is rational. A person is seen as not all good or all bad, but as sometimes good, sometimes bad, sometimes in between. A viewpoint is seen as neither right or wrong, but as being right in some ways, wrong in others, and "maybe right, maybe not" in others.

But awareness is limited. The individual views events from *one* perspective—usually his own. He acts to achieve *his* goals, without considering the viewpoints of others. If another viewpoint differs from his, he doesn't understand. He may try to "reason" with the other person to show him the "error" of his ways. When this does not succeed (as it seldom does!), he may become impatient or even angry. He may conclude that the other is "dumb" or "stupid."

The Uniordinal Mode is the characteristic mode of most normal human beings. By "normal" we mean "free of serious mental disorders." It is a mode of *limited rationality.* Unless this mode is transcended by a "critical mass" of human beings, the future of the human race is not very bright.

Multiordinal Mode. As its name implies, this is the mode in which the mind views events from two or more perspectives at once. The individual acts to achieve his own goal of the moment, but is at the same time aware of the relationship of his action to other possible goals or viewpoints. In a situation where his action affects or may affect another, he considers the other's viewpoint. He is able to "put himself in the other fellow's shoes." He doesn't abandon his own viewpoint of the moment. But he is ready to modify his action to accommodate that viewpoint wherever he can do so without undue sacrifice of his own goals or interests.

By viewing an event or situation from two or more viewpoints, the individual often sees that a given action, if done in a certain way, can promote both viewpoints. His actions thus tend to be multipurposed.

If something happens that is relevant to a viewpoint other than his immediate goal, he quickly sees this. Sometimes this relevance to a second viewpoint is more important than the immediate goal. In science, this is called serendipity—the ability to see the significance or value of an unexpected finding. A person who is uniordinal in mode would fail to see this.

Synergic Mode. This is a natural development of the Multiordinal Mode. Clearly, there is a limit to the number of viewpoints a person can hold simultaneously since the human mind is finite. Furthermore, a person can become so preoccupied with finding multiple perspectives that he gets hung up on multiplicity, so to speak, and gets nothing done.

In the Synergic Mode, the mind is multiordinal, but it is also focused sharply on things that promote two or more viewpoints, goals, perspectives, etc., while impeding none. In addition, the mind is especially alert to basic viewpoints that embrace several other perspectives as "special cases." This sharp, clear focus on *synergies* as they emerge quickens mental function beautifully.

But there is more to the Synergic Mode than this. As synergies occur more frequently, a new level of integration of the mind begins to emerge, a holistic level that is difficult to describe in words. So much goes on so fast! But more on this later.

The synergic mind is *sane,* in the sense used by Korzybski.(1) It clearly distinguishes between perceptual and verbal "maps" of reality and the "territory" those maps represent. There is continuous cognizance that verbal maps are often inaccurate in some ways and almost always incomplete. There is continuous correction and instant adjustment of these maps in accordance with new data. The distortions of wishful thinking and fearful thinking are eliminated.

The synergic mind is *rational* in the highest and best sense of the word. It does not use reason to justify beliefs already held. It uses reason to examine these beliefs in order to determine the consequences. It follows reason wherever it leads. It does not fear new ideas or shrink from strange concepts.

The synergic mind is *ethical.* It naturally thinks not only in terms of its own viewpoint and interests, but those of others as well. It selects its own goals and guides its actions so as to promote the interests* and goals of others, or at least so as not to impede them.

Indeed, this high ethics quotient (E.Q.) of the Synergic Mode is one of its most beautiful, remarkable and exciting characteristics. It is impossible for a synergic mind to perform an unethical action and remain synergic. This seems incredible, and it does not mean that a person in the Synergic Mode is incapable of such an act. What it means is that if he does so act, he loses the Synergic Mode.

There is much more to the Synergic Mode than is described here. Many readers, I am sure, have encountered this phenomenon at some time in their lives and therefore know in concrete terms what I mean. What especially turns me on is the tremendous potential of the Synergic Mode; I have, so to speak, "gone synergic" on it. *For the*

---

*This needs to be clarified. The synergic mind is capable of recognizing when the goal of another is reactive, or when his interests are dysergic. It does not support a reactive goal or a dysergic interest. But it also makes a distinction between such goals and interests and the *person* driven by them and supports the *person* even while rejecting the dysergy. A person driven by dysergy is in pain. The synergic mind "reads" beyond the dysergy and finds a way to soothe the pain.

Synergic Mode is available to any human mind. It is one of the sublime prerogatives of being human. And it seems clear that if more people used the Synergic Mode more of the time, the world would be a better place.

This, then is the Mode Ladder. It is the basic yardstick of synergetics. It is possible, using this yardstick, to estimate the mode of another person or to evaluate your own mode at a given moment. This yardstick is not, of course, as precise as those of the physicist or engineer, but it is very useful. With a little practice, one can make very shrewd estimates of the mode of a mind, or the mode of a group, or the mode of any functioning system. Knowing the mode, a syngeneer* knows what to do and what not to do—he suits his action to the mode.

Movement up the Mode Ladder is called *traverse* in synergetics. In terms of the Mode Ladder we can formulate the basic goal of synergetics very precisely: *Evoke traverse.*

---

*The noun "syngeneer" is a coined word meaning one who generates energy. The verb "to syngeneer" means to change a situation in a more synergic direction.

The sharp separation of the descriptive from the normative is the heritage of modern empiricism... This world view holds that the world is predominantly empirical and can best be rendered by physical science, whose ultimate conceptual components are descriptions, or scientific explanations. Within the consistent scientific world view, norms and values are relegated to limbo. They have no genuine cognitive status... The ideology of modern science, which itself is a value system, so interprets other value systems as to make them either unimportant or meaningless.

Let us first of all notice that description of what is out there largely depends on what we assume to be there... We always have to observe something in order to determine something. The myth of pure observation is a part of the Baconian heritage. We have now reconciled ourselves to the fact that there is no such thing as pure observation. Every observation is a directed observation, is an observation for, or against, a point of view.

——Henryk Skolimowski (1)

CHAPTER 3

THE HEURISMS OF SYNERGETICS

As every schoolboy knows,* geometry is based on a number of fundamental statements called axioms and postulates. Indeed, all of mathematics is so based. These were once regarded as self-evident truths, not requiring proof; more recently they tend to be regarded simply as basic premises, the truth of which is assumed and from which the truths and theorems of mathematics are logically derived.

Mathematics is one of the most remarkable achievements of the human mind. The elaborate and beautiful structure of ideas that mathematicians have developed is truly awe-inspiring. Their success has led workers in other fields to emulate them, with results that have not always been as impressive.

---

*Or used to know! The "new math" does not appear to have changed this, however.

30

Physical science is very much like mathematics and indeed uses mathematics as a language and tool for analysis. Physical theories such as quantum mechanics often have an axiomatic foundation. But physical science differs from mathematics in that its theories are subject to the discipline of experiment. A physical law is a general statement to which no exceptions have yet been observed. If an exception is found, the law is abandoned, and a new, more general statement is derived. Again, the success of physical science has inspired emulation, which often has not been equally successful.

In dealing with the human mind, or with systems involving human beings and groups, the paradigm of physical science has one serious defect: there is no place in it for human values. Moreover, it tacitly assumes that a human being is a passive entity whose behavior is entirely controlled by laws that can be observed. But a human being is not entirely passive; he is also an active agent who can observe that he is being observed and respond out of pure devilment in unpredictable ways just to confound the observer. And a human being does have values and purposes. This inevitably makes a procrustean bed of the physical science paradigm when it is applied to man.

I hope the reader will pardon this obstruse excursion into the labyrinths of epistemology; it was done for a purpose. The purpose is to provide some explanation for the somewhat unorthodox approach to a *science* of synergy that I believe to be necessary. I love mathematics and physics and indeed have devoted many hours to their application to biomedical problems; they definitely have a role to play in biology and medicine, and indeed their role is even today inadequately appreciated. But something more is needed where the human mind is concerned.

This has led to the concept of the heurism* as playing a role in synergetics analogous to that of the axiom in mathematics. Synergetics is based on an evolving set of heurisms. These may be regarded as synergic self-fulfilling prophecies. As proposed by Merton (2), a self-fulfilling prophecy is a belief regarding a social situation which leads an individual or group to act as if the belief were true, with the result that other individuals or groups affected by such action also respond as if the belief were true, thus reinforcing the belief. Thus, if group A believes group B to be hostile, group A may act in such a way that group B responds in a hostile way, even though it was not hostile to begin with.

---

*A similar concept, called a *heuristic assumption,* is sometimes used in sociology. (3)

A synergetic heurism is a general principle or proposition which, when accepted as a basis for action, *leads to the emergence of synergy*. The proposition itself can neither be proved nor disproved; facts against it can be marshalled as readily as facts in support of it. A decisive experiment cannot even in principle be designed to prove a heurism true or false, because one or more human beings are involved. If they know of the heurism being tested, they can interpret it as true or false and act accordingly. If they do not know of the heurism, the experiment is not decisive, since it omits a vital component of the system being tested. A synergetic heurism has the additional property of being a synergic self-fulfilling prophecy, i.e., one that conduces to the emergence of synergy. Thus, its adoption as a basis for action tends to evoke synergy—synergy that might not otherwise occur.

Here, then, are the heurisms of synergetics thus far adopted.

1. Analytical Principle. *Optimum development of the rational (analytical) mind of a human being will occur as a result of the rational application of knowledge about that mind to its own development.*

This may appear to be a somewhat roundabout way of stating the old Socratic maxim: "Know thyself." It asserts that knowledge about a human mind is of key importance to its development. But knowledge alone is not enough; it must be rationally applied for the purpose of developing that mind. There must, in other words, be synergy of knowledge and application.

As stated, the Analytical Principle may sound rather abstract on the one hand, and perhaps rather obvious on the other. But it has a very important practical consequence, which may be stated as a corollary: *the individual's own rational mind can solve all its own problems.* If a problem is not solvable as stated, the rational mind discerns this and restates the problem in solvable form. This heurism and its corollary are basic to many of the synergetic techniques described in this book.

The Analytical Principle is also basic to a process of "bootstrapping" a human mind. The more a mind knows itself, the better is it in a position to apply that knowledge to its own development. The more such an application is done, the greater the capacity of a mind for knowing itself. Thus, another corollary of the Analytical Principle is: *the human mind has the capacity to transcend itself and to do this repeatedly.*

This heurism, like the other heurisms of synergetics, is basic and general; it has a potentially infinite variety of applications. Some of these will become evident in later chapters of this book. At this time, the reader may wish simply to meditate for a while on this principle, letting it filter through the many layers of his mind.

2. Synergic Principle. *Synergy of function is promoted by regarding all processes of experience as potentially of value to the individual.*

In other words, anything that happens can be useful. This means, for example, that even so-called "bad" emotions such as greed, lust, hate, envy are potentially of value. They are the repressed and un- developed indicators of potentially new abilities.

An advantage of this principle is that it establishes a basis for integration. *No* part of a human being need be *absolutely* condemned.

If a part of an individual is rejected by other individuals, or rejected by himself, then he may adopt a belief that he is "bad" or "inferior." The Synergic Principle provides a new outlook, enabling a person to accept these "bad" parts and to initiate a process of development. Sooner or later, he *changes.*

Again, it is suggested that the reader meditate on this principle for awhile.

3. Staging Principle (Principle of Metacognition). *Any work method requires some basis, or stage, upon which to proceed. The stage chosen exerts a profound influence on the phenomena that are produced, and all such phenomena will tend to take a form that fits the stage chosen.*

This means that any approach to human development works—up to a point. If a person is studying yoga and practices the techniques of yoga, the resulting phenomena will tend to arrange themselves to conform to the basic teachings or assumptions of yoga. If a person decides to undertake psychoanalysis, or transactional analysis, or encounter group training, or synergetics, a similar process tends to occur. The stage determines the form in which phenomena appear.

But it also follows that there is no universal stage. Furthermore, a given stage may be optimum for one person at a particular phase of his development, but not for another.

From a heuristic standpoint, this principle suggests the concept of *multiple staging:* the use of several stages rather than a single stage as a basis for a work effort. Since only in rare instances will a single stage be optimum for an individual, the use of several different stages is likely to be more effective in the long run than confinement to a single stage.

The effectiveness of multiple staging is enhanced when the several stages are *synergic* with one another. The various work methods of synergetics have been designed with this in mind.

4. Absolute Uncertainty Principle. *The only absolute certainty is that there are no absolute certainties.**

A scientific law or a mathematical theory, in the light of this principle, may be regarded as having a high probable truth value, say, .9999998, where 1 represents absolute certainty (never achieved). Suppose a proposition is regarded as absolutely certain by a particular mind. That mind is closed permanently as far as that proposition is concerned. If it later turns out that the proposition is not true in some cases, such a mind will be led into error; worse, it will be unable to detect or correct the error. On the other hand, if the proposition is accepted as having a high probable truth value by a mind, that mind will be able to deal adequately with the occasional error. This is an extreme situation. The more usual one is where a mind accepts a proposition as absolutely certain when in reality it has only a moderately high probable truth value.

The synergic value of the Absolute Uncertainty Principle is that it keeps a mind always basically open, always searching, always evolving.

5. Principle of Synergic Efficacy. *In the long run, a synergic approach to a problem or situation involving humans is more effective than a dysergic approach.*

A dysergic approach may gain a temporary advantage; this is one reason why such approaches are so often used. But a dysergic approach sets in motion counterprocesses that must be dealt with sooner or later.

A synergic approach on the other hand, may sometimes be slower or require more immediate effort; but it does not evoke counterprocesses.

When an individual or group is reactive in mode, decisions or actions are usually dysergic and set in motion counterreactions by those affected. A syngeneer may on occasion be reactive in mode; but

---

*This has deliberately been stated in paradoxical form for dramatic emphasis. A more exact statement would be: The only absolute certainty is that there are no absolute certainties, except the absolute uncertainty principle.

if instead of reacting and producing dysergy, he first of all traverses to synergy and then deals with the problem or situation, his action is usually more effective in the short run and does not evoke counter-reactions, which would otherwise tie up time, effort and thought in the long run.

An exception to this principle should be noted. In an emergency, actions that evoke counterreactions may be taken, which are necessary under the circumstances. Even here, however, the principle applies: after the emergency is over, steps may be taken to eliminate the dysergy that has been produced and to promote synergy.

6. Principle of the Natural Synergy of Man. *Human beings are naturally and basically synergic.*

The existence of dysergy in human individuals and groups is not denied, but it is explained as being caused by *acquired patterns* that are not an essential or basic part of human nature.

To put it differently: synergetics belongs to those schools that hold that man is basically "good"; that "evil" in man is something that is acquired and can therefore be eliminated.

Admittedly, this proposition can neither be proved nor disproved; and there are individuals and situations to which it can be applied only with difficulty. But if the principle is rejected, or if its opposite is accepted, the possibility of synergic development of humankind is foreclosed. It is like going into a game convinced you are going to lose; what's the use, then, of trying?

Acceptance of the principle, on the other hand, does not mean a denial of the existence of dysergy; but it always keeps open the door to the possibility of eradicating dysergy and achieving synergy. In particular, it is exemplified in the remarkable finding that a person functioning in the synergic mode is naturally ethical; that if he at any time acts unethically, he inevitably experiences a "mode drop."

7. Principle of Synergic Grasp of Synergetics. *Application of synergetic ideas and tools by a person or group is most effective when the person or group has a synergic grasp of those ideas and tools.*

The person who has an excellent grasp of theory but is all but-terfingers in practice is well-known to all of us. Conversely, a person may know how to fix things or to make things or even to run things, but without any real understanding of what he is doing or why. In other words, there are at least two kinds of knowledge: conceptual grasp and practical know-how.

But there is a third kind of knowledge available that results from a *synergic combination* of conceptual grasp and know-how. This is called *synergic grasp;* and it constitutes a whole that is greater than the sum of its parts.

One emergent of this synergic whole is the phenomenon of *think-feel synergy*. When a person has synergic grasp of an idea or tool, he becomes imbued with an extra-awareness—an awareness of the enormous range of potential applications of the idea or tool—a range that far exceeds the immediate purpose for which it was designed. Accompanying this awareness, there is an appreciation of the tremendous potential value of the idea or tool, which in turn fills the individual with enthusiasm—what Pasteur called the "inner God."

A synergic grasp of synergetics will clearly enhance the effectiveness with which it is used.

These, then, are the basic heurisms of synergetics, as it has thus far evolved. They are stated in abstract terms whose relevance to concrete human problems and situations is not immediately evident. They do, however, provide a foundation upon which the structure of ideas and tools described in the rest of this book has been erected. They also provide a basis for the application of synergetics to other fields of human endeavor. Furthermore, in any particular field of application, problems sometimes arise that are not manageable with the existing set of ideas and tools. Such problems require the creation and development of new ideas and tools. These heurisms are useful in such situations.

Finally, it should be emphasized that other synergetic heurisms are possible, which the reader may wish to introduce. This is encouraged, the only suggestion being that care be taken to ensure that a new heurism be synergic with those given here.

For convenience, the heurisms are repeated below:

1. *Analytical Principle.* Optimum development of the rational (analytical) mind of a human being will occur as a result of the rational application of knowledge about that mind to its own development.
2. *Synergic Principle.* Synergy of function is promoted by regarding all processes of experience as potentially of value to the individual.
3. *Staging Principle.* Any work method requires some basis, or stage, upon which to proceed. The stage chosen exerts a profound influence on the phenomena that are produced, and all

such phenomena will tend to take a form that fits the stage chosen.

4. *Absolute Uncertainty Principle.* The only absolute certainty is that there are no absolute certainties, except the Absolute Uncertainty Principle.

5. *Principle of Synergic Efficacy.* In the long run, a synergic approach to a problem or situation involving humans is more effective than a dysergic approach.

6. *Principle of the Natural Synergy of Man.* Human beings are naturally and basically synergic.

7. *Principle of Synergic Grasp of Synergetics.* Application of synergetic ideas and tools by a person or group is most effective when the person or group has a synergic grasp of those ideas and tools.

There will come a time, I know, when people will take
delight in one another, when each will be a star to the
other, and when each will listen to his fellowman as to
music. The free men will walk upon the earth, men great
in their freedom. They will walk with open hearts, and
the heart of each will be pure of envy and greed, and
therefore all mankind will be without malice and there
will be nothing to divorce the heart from reason. Then
life will be one great service to man! His figure will be
raised to lofty heights—for to free men all heights are
attainable. Then we shall live in truth and freedom and
in beauty, and those will be accounted the best who will
the more widely embrace the world with their hearts, and
whose love of it will be the profoundest; for in them is the
greatest beauty. Then will life be great, and the people
will be great who live that life.

——Maxim Gorky

CHAPTER 4

OVERVIEW—
THE SYNERGETIC PROGRAM

In previous chapters, the Mode Ladder was described, and the basic goal of synergetics was presented: to evoke traverse up the Mode Ladder to the synergic mode of function. The heurisms of synergetics were then described—basic principles that have guided the formulation, development, testing, and evaluation of synergetic ideas and tools. In this chapter, we complete the presentation of Basic Synergetics as it has evolved to date, with a description of another basic concept—the Status Cross—and an overview of the rest of this book.

It is pretty obvious that men are not created equal. The idea of equality, taken from a literal interpretation of the words of the American Declaration of Independence, can lead to some rather difficult ideological positions.

But Thomas Jefferson did not mean that men and women were

38

identically equal. He meant that, in a just social order, all persons should be *treated* equally, that none should have special privilege by virtue of accident of birth, wealth, or social position. Each person is a unique individual; but each is entitled by inalienable right to equal protection of the law, to equal treatment by the law, and to equality of economic and social opportunity. Only when all men and women have social equality* can the unique potential of each be realized, for the ultimate benefit of all.

In synergetics, we formulate this idea as the Principle of Equivalence of Status. This may be stated as follows: the flow of synergy, empathy, and communication between two individuals is optimum when they have equivalent status with respect to each other. Referring to the figure below, equivalent status is indicated by the center of the cross.†

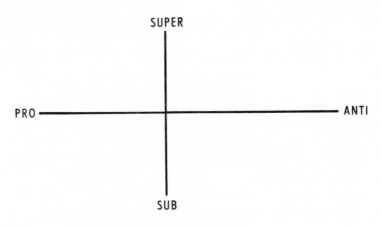

THE STATUS CROSS

When one regards the other as Super, the flow of synergy, empathy, and communication (SEC) tends to go down.

When one regards the other as Sub, SEC goes down.

When one regards the other as Pro, as someone to be dependent upon, SEC goes down.

When one regards the other as Anti, SEC goes down.

---

*Even here, social equality does not mean *identical* treatment, merely that there is some kind of fair balance—in a word, synergic equality.

†Actually, the optimum position may be considered to be slightly to the left and above the center—i.e., when each regards the other as somewhat Pro and somewhat Super.

The Status Cross is of basic importance in Group Synergetics. It is useful in analyzing dysergic group processes. For example, if communication is poor between two members or cliques in a group, it may be found that they do not occupy the Equivalence of Status point on the Status Cross. Corrective action to establish Equivalence of Status will then often lead to improved two-way communication. From a positive viewpoint, group synergy can achieve remarkable heights when a group is so organized that Equivalence of Status is maintained.

Two special applications deserve mention. In writing this book, I have collected into an organized whole a set of ideas and tools which, in my studies and experience, demonstrably promote synergy and/or reduce dysergy. But please don't make me a guru. These ideas and tools should be independently tested and evaluated by each reader—some may be in error, others are no doubt poorly stated. Furthermore, the experience of the reader doubtless includes data not available to me, about ideas and tools that promote synergy and reduce dysergy. Synergetics is not what the artcoulter says it is; it consists of ideas and tools which, in the collective experience of syngeneers, have been effective in promoting synergy and reducing dysergy.

The second application is in the domain of work methods, especially in the roles of the coach and the group monitor (discussed in more detail in later chapters). It is natural for a coach or a group monitor to be accorded a Super or a Pro status. (He also not infrequently acquires a Sub or an Anti status!) It is important that steps be taken to prevent this or to correct it whenever it occurs. *Synergetics works far better when Equivalence of Status is maintained.*

It is, of course, true that an experienced, well-trained, knowledgeable coach or group monitor can achieve more than an inexperienced or inadequately trained one. But this is all the more reason for him to operate from an Equivalence of Status position. If he's really good, he doesn't need a Super status; and if he can't operate from a position of Equivalence of Status, he's really not as good as he may think he is.

Basic Synergetics, then, as it has thus far evolved, consists of the Mode Ladder, the synergetic heurisms, and the Status Cross. Many more ideas and tools could be included, but I have deliberately kept this as simple as possible. Starting with Basic Synergetics, a creative syngeneer can readily formulate, develop, test, and apply other ideas and tools appropriate to other fields not covered in this book. And the ideas and tools of Basic Synergetics are used, directly and indirectly, in all the branches of Synergetics covered in this book.

Let us now consider the synergetic program. The basic goal of synergetics is to promote synergy and/or reduce dysergy in functioning systems. In this book, we apply this goal to individuals, to groups, and to communities. Part Two deals with Individual Synergetics, Part Three with Group Synergetics, and Part Four with Social Synergetics.

It should be emphasized that these three parts are not isolated from one another, but form an interrelated whole. Desirably, they should constitute a synergic whole. However, it is necessary to take them one at a time. This inevitably means that, on the first time through, the reader will get only a partial and inadequate picture. The reader who wishes to acquire a synergic grasp of synergetics, therefore, is advised to read the rest of the book twice. The first reading will gradually give him a picture of the whole. The second reading will enable him to place each detail in synergic relationship to other details and to the whole.

Even this will not provide full synergic grasp. In addition, experience with the concrete phenomena that are evoked when these tools are used is necessary.

Individual Synergetics consists of two main parts: ideas and tools for promoting synergy, and ideas and tools for reducing or eliminating dysergy. The two are not isolated, although considered separately. The production of synergy makes it easier to eliminate dysergy. The elimination of dysergy makes it easier to achieve the synergic mode. The two efforts thus reinforce each other.

Basic to Individual Synergetics is the synergetic *session*. This is analogous to, but distinct from, the production of special states of consciousness in yoga, in psychoanalysis, in hypnosis, etc. In psychoanalysis, for example, the patient lies on a couch and free associates, letting whatever comes to mind emerge without censorship and reporting verbally to the analyst. In a synergetic session, a technique called *tracking* is used. This induces a state of enhanced rationality in the tracker, occasionally punctuated by periods in which the synergic mode turns on. Many other techniques may also be used, but tracking is the basic technique.

A synergetic session may be carried out by an individual working alone or with the help of a friend who acts as a *coach*. In Part Two, it is assumed that the tracker is working alone. Tracking, however, is a skill that is not acquired without effort; it takes practice and an investment of time, energy, and thought. A coach who knows how to track can be very helpful to a person who is learning to track. Paradoxically, a competent tracker can also use a coach to assist him and often achieve more than he can working alone.

Coaching is also a skill that requires an investment of time, energy, and thought. It is defined and governed by a set of principles and policies called the Coach's Guide. Although superficially similar to some forms of psychotherapy, it is *not* psychotherapy. This point needs to be emphasized. The basic goal of coaching is different and the *staging** is different from that of psychotherapy.

Just as Individual Synergetics promotes synergy and reduces dysergy in individuals, so does Group Synergetics promote synergy and reduce dysergy in groups. Group Tracking is the basic technique. Desirably, each member of a group should have learned how to track, or at least be learning how, in order for Group Tracking to be effective. Conversely, an individual can usually make better progress in Individual Synergetics if he concomitantly is a member of a synergetic group.

One remarkable and exhilarating feature of Group Synergetics is a special type of group called the Synergic Team. A Synergic Team is a group functioning in the synergic mode as a group. A wonderful spirit of mutual trust, love, and admiration emerges, combined with a synergy of interaction in thought and deed that has to be experienced to be understood. Some athletic teams develop a similar relationship, but perhaps not with the same degree of mental synergy. Being part of a Synergic Team is one of the most rewarding aspects of synergetics.

Part Four deals with Social Synergetics, a field pioneered by Lester Ward and Ruth Benedict and developed considerably in recent years by James and Marguerite Craig, Abraham Maslow, Walter and Nancy Strode, Wes Thomas, Donald Benson, Sadah Loomis, and the Committee for the Future, originators of the SYNCON type of conference, to mention only a few.† Potentially, this is the most important single branch of synergetics. Most would agree, I think, that our strife-wracked planet could use a little more synergy.

The basic program of synergetics has these goals: to develop

---

*Recall the Staging Principle discussed in chapter 3.

†These are persons who explicitly have applied the idea of synergy to social processes. Many others have presented ideas and tools that promote synergy and/or reduce dysergy in social processes, without explicit reference to the synergy involved.

synergic individuals, synergic teams, synergic communities, and ultimately a synergic world order. Any objective, realistic appraisal of the condition of humankind quickly leads one to conclude that we have a very, very long way to go before these goals can be achieved. But humanity is young, as species go; in a short few thousand years humans have progressed at an amazing rate and have transformed the face of the earth. True, not all progress has been beneficial, and we now face a number of crises any one of which could lead to catastrophe if not resolved. But a species that has achieved so much in so short a time span should not be underestimated. We believe in humanity. We believe in the essential goodness of human beings. And we believe in the electrifying power of synergy—a new force in human affairs.

PART TWO

Individual
Synergetics

Over the last hundred years, we have frequently seen the statements in many different guises that Copernicus dethroned man as the center of the universe, that Darwin removed his claim to uniqueness, that Freud dispossessed him as master in his own body. Nowadays, the resigned attitude prevails that computers have removed the necessity for his very existence as far as thinking is concerned.

I maintain that the sum total of the mathematical analysis of the processes of control and decision-making carried out over the last twenty years conclusively refutes this pessimism. It makes evident that the human mind is an instrument of fantastic power and subtlety, far beyond our abilities to comprehend. Its powers are barely tapped, and it is impossible to predict its growth over the coming centuries, aided and reinforced by technical aids such as computers. Whatever happens, the human mind will remain dominant and in the central position. The human spirit remains far above anything that can be mechanized.

——Richard Bellman(1)

A human being is, so to speak, a continuously operating push button of change. All the amplifications and effects that can be produced by human interaction are to some degree at the command of each of us. . . We live in a billionfold kaleidoscope of infinite potentialities, changing at every moment in its causal details, with many of the changes hidden in these billionfolded selecting and amplifying heads of ours. . . Every moment branches out into a vast and unpredictable future that you are changing just by reading this, or daydreaming, or blinking your eyes.

——John R. Platt (2)

CHAPTER 5

THE HUMAN POTENTIAL

Every human being is unique. We are not mass-produced according to some blueprint or master plan, each identical with the other. Each of us emerges from a different design, a different set of genes. But more than this, each of us has a unique history—a unique sequence of events that happened to us, together with our responses to those events and our reflections on the experience. Even identical twins, having duplicate genes, are distinguished from each other by their unique histories. In short, you are one of a kind. There is no other person in the world quite like you, there never has been, and there never will be.

In addition to being unique, every human being is precious. It took a billion or more years of evolution to make you what you are—an evolutionary process that is itself unique. Moreover, you are a being of incredible complexity—the design of the human ear or the human eye, for example, is simply magnificent. As for the human brain, it is a supercomputer whose intricacies and powers are far, far in advance of any of the artificial computers, which simply imitate and expand the simplest of those powers. The fact that a computer can do arithmetic much faster than a human brain may be of interest, but the really remarkable fact is that human brains invented arithmetic and designed computers to do it.

All this implies that each human being has a unique potential and that it is simply outrageous that everything possible is not done to permit that potential to develop. Yet every society on this planet not only does not do this, but is full of barriers and pressures to prevent it!

There are, for example, the twig-benders. These are groups that consider children to be a form of plant life and seek to capture them at an early age, hoping to bend the twigs in a direction that will force them to grow the way the twig-benders want them to. And so they indoctrinate them with their TRUTHS and inculcate them with their VALUES and above all instill in them habits and attitudes to ensure their obedience and conformity.

Of course, it doesn't work. Children aren't twigs. They are self-determined beings with a sense of their own individuality and worth, and they naturally rebel. But they are also small and dependent and, to the degree necessary, they submit to the twig-benders. The result is not only a messed-up world with a lot of messed-up people; far worse than that, it is a tragic waste of human potential. To paraphrase the poet, we all end up strangers and afraid, in a world we never made. And the tremendous potentials of our unique minds remain undeveloped. Comparatively speaking, we are mental dwarfs when we could have been giants.

Individual Synergetics starts with the heurism that we are unique, self-determined beings. Unlike some schools, its goal is not to eliminate "aberrations" or "neuroses" that cause people to deviate from "normality" (whatever *that* is), but to provide ideas and tools to enable the individual to eliminate the impedances blocking his unique development and to activate the unique synergies of his own mind. That is why we insist that the individual is always in charge of his own case. That is why we insist that coaching is not a form of psychotherapy,* which implies that the coach is an Authority who Knows Best. No other person, no matter how wise or clever he may be, no matter how many books he has written or degrees he has earned or patients he has treated, can possibly know your mind as well as you do. True, he may see things that you have blocked from your awareness; but his vision is always partial and incomplete and superficial, from the outside. You are the only one who can see your mind from the inside; you are the only one who has access to all the data; you are the only one who can fit all the pieces together into a synergic whole.

It is this uniqueness that we respectfully and lovingly address; and all the ideas and tools of synergetics—no matter how pedantically they may be expressed—are designed from this perspective. Try them out if you wish; use them if they work; but never hesitate to adapt the tool to *your* needs or to change it to fit your own knowledge and experience.

The first step in Individual Synergetics—and the foundation of all that follows—is to focus on your uniqueness and to take charge of your own development. From this perspective, let us now examine the human potential, bearing in mind that everything that is said needs to be modified and tailored to fit that wonderful uniqueness.

The idea that the human mind is "an instrument of fantastic power and subtlety" whose "powers are barely tapped" has occurred to many human minds at various times and places. It is an appealing idea. Everyone would like to be supersmart, have total recall, and be irresistible to the opposite sex. The very appeal of the idea leads us to be defensive about it. At the same time, charlatans constantly exploit this appeal for their own enrichment, making matters more difficult for serious workers in this field. Despite these handicaps, there has

---

*This is not said in denigration of psychotherapy; indeed some schools such as Rogers's client-centered therapy have similar approaches, and many psychotherapists, in practice, have an outlook not very different from ours. Some people are, through no fault of their own, in need of the special kind of help that psychotherapy provides. In a sense, for these people, synergetics begins where psychotherapy leaves off.

been growing interest recently in the development of the human potential. (3)

No attempt will be made in this chapter to present a detailed chart of the potential abilities of the human mind. Instead, I will simply outline some domains of experience and action that are available to humans but do not appear to be fully used. These domains are used extensively in synergetics.

The average person appears to function largely on what we call the *mind band* of experience—he identifies with his ordinary consciousness and will. There is, however, potentially available an expanded consciousness, which we will call the *broad band*. Just as the discovery of the electromagnetic spectrum made possible a host of new inventions such as radio, television, x-rays, infrared lamps and cameras, so may the exploration and use of the broad band make available new abilities to the individual.

It is convenient to describe the broad band in terms of the following domains:

1. The tempos
2. The tracks
3. The holistic level
4. Synergic team functions

The Tempos

Whatever the ego is aware of, at any given moment, we call the *contents* of consciousness. A content may be a sensation—the sight of a tree, the sound of a passing car, the smell of a rose, the taste of orange juice. Or it may be an idea—the idea of justice or conformity or happiness. It may be a mental image—the face of a loved one or a barber pole or a prune. It may be an emotional feeling of sadness or excitement or fear. It may be the recollection of a past incident or the anticipation of a future event. At any given moment, a large number of contents present themselves to awareness. The ego can selectively focus attention on some contents while ignoring others; but the focus is ordinarily on contents.

These contents are usually not fixed or static, however. As time goes on, they may change—in location, in intensity, in the features they present, in their relations to other contents. They may disappear from awareness while new contents appear. The continuous shifting, changing, emergence and disappearance of contents was described by William James in a famous metaphor as the *stream of consciousness.*

The stream of consciousness may be regarded as composed of a number of *processes* involving the various contents. Now, just as physical objects in motion have different velocities, so do the processes of the stream of consciousness have different tempos. Some occur very rapidly, others slowly, some so slowly they appear to be stationary. It is convenient to select one process whose characteristic tempo is familiar to all as a basis for comparison—the process of ordinary speech. We refer to processes having the tempo of speech as *orthoprocesses.* Those that occur much more rapidly, we call *microprocesses;* those that occur much more slowly, *macroprocesses.*

We could, of course, devise a *spectrum* of process tempos, analogous to the electromagnetic spectrum. But it is unnecessary to be so precise for our purposes. Just as visible light is used as a reference band in the electromagnetic spectrum, with infrared on one side and ultraviolet on the other, so can we roughly identify whether a process is an ortho, micro, or macroprocess.

It is at once evident that in ordinary consciousness, attention is focused almost entirely on orthoprocesses. Yet we can, if we choose, examine micro and macroprocesses. Indeed, *this is one way consciousness may be expanded.* We are not here referring, it should be noted, to the alteration of time sense that occurs under the influence of certain drugs, or in the state of hypnosis. What we refer to is a different kind of consciousness-expansion, one which opens the way to the development of a number of "new" abilities.*

The microprocesses are particularly interesting. Usually we are unaware of their existence, but under special conditions we realize that an extraordinary number of very fast processes go on in company with the slower orthoprocesses.

A man driving a car with casual control suddenly observes the cars ahead abruptly stopping. In a flash he (a) evaluates his own speed, (b) predicts he cannot brake in time to avoid a collision, (c) evaluates the left lane to be unsafe, (d) decides to swerve to the right onto the shoulder, and (e) does so. All these processes occur in a fraction of a second. The driver may not be fully aware of them at the time, but they are recorded in memory and he can readily recall them. They occur in a rapid-fire sequence of flashes. In this case, they are simply

---

*Strictly speaking, these abilities are not "new"; we've always had them. They are, however, *undeveloped*, since our main focus of attention is on orthoprocesses and is constrained to that tempo in social discourse.

processes that ordinarily occur at ortho tempo, but have been speeded up under stress. (They are not "instinctive" because another person might panic under the same circumstances, and each process is one previously learned by the driver.) Not all microprocesses are of this type, however (i.e., speeding up of orthoprocesses).

Microprocesses occur frequently when the synergic mode is turned on, and indeed are one of the delights of the synergic mode. The experience of thoughts racing along several tracks simultaneously can be highly exhilarating. The expansion of consciousness to include microprocesses in addition to orthoprocesses is well worth the effort, in our opinion.

On the other side of the orthoprocesses are the *macroprocesses*— processes that go on so slowly that they usually escape notice, except for the vague realization that things have somehow *changed*. But they are there, and they are every bit as interesting as the microprocesses.

The mind-dweller* characteristically is occupied with the present. He bases his judgments on the perspective of the moment and shifts with the tide as it turns without being aware that the tide is there. He interprets the past purely in terms of the values of "now," and anticipates the future in the same terms.

Yet macroprocesses do occur; and most of us use them and are aware of them, to a degree. The flexible, patient pursuit of a long-range goal; the consistent application of a policy; the follow-through on a decision—these are examples of processes that occur at slow tempo that are familiar. However, there are others that go on that escape our notice, for which we have no names. We may look at a problem today and feel there is no way to solve it; the next day, looking at the same problem, we suddenly see how easy it is. The problem did not change, we did; yet we are unaware of the process by which this change occurred. This is another example of a macroprocess. We can expand our consciousness so as to become aware of these processes and to develop new abilities that use them.

To see each present moment in itself, in all the boundless variety and richness there to be found is, of course, important.† But one can

---

*A term sometimes used, with affectionate respect, for the person who focuses his attention entirely on orthoprocesses (and Main Track, described below). No derogation is intended; we are all mind-dwellers in various degrees. Conversely, no one is purely a mind-dweller; we have set up a straw man to aid communication.

†There are schools that emphasize present-time orientation, stressing the *here* and *now* as all-important. Where this is done to free individuals who carry the past around with them like a dead weight, or who are preoccupied with worries about the future, such an emphasis is useful. But a present-time orientation can also be overemphasized, to the extent that it shuts off awareness of macroprocesses.

do this without being *stuck* in present time. The domain of macroprocesses can also be lived in; it enables one, so to speak, to function as a four-dimensional being, to whom each "now" is but a phase of a process flowing on, and in terms of a perspective from which all "nows" are "present."

It is in this domain that an individual *evolves*. These are the processes by which we may effect lasting changes in our being. They provide the means for achieving temporal organization of our experience. It is a domain well worth knowing better and using more.

The Tracks

As mind-dwellers, we not only confine ourselves to the *tempo* of orthoprocesses; we also limit our orientation to the contents of consciousness—the sights, the sounds, the images, the feelings, the desires, the memories, and so on.

But these contents do not just happen; they are produced by an activity that we refer to as *operations*. Thus, we *associate* one idea with another; we *compare* these ideas, noting similarities and differences; we *recall* a memory of a previous incident; we *search* for a felt idea; we *express* or *sublimate* or *repress* an emotion. Each of these acts is an *operation*, and, of course, we have always known of their existence. But our characteristic orientation is toward *contents*, not operations.

We may make the distinction between content and operation clearer by comparing what goes on in our minds with what goes on in a movie. The *contents* of awareness are like the moving picture on the screen; our attention is focused on the screen. The operations of awareness are like the processes going on in the movie projector. We rarely pay any attention to the projector.

Yet there is a simple act by which we can shift our orientation from *content* to *operation*. Curiously, this act apparently has no name. Borrowing a term from electrical engineering, we refer to this act as *phase shift*, because it is a shift in the phase of orientation.* Phase shift goes counter to the "natural tendency of the mind"; but it is a simple act and one that is readily learned. With practice, it is possible

---

*Phase shift is not the same as introspection, which is defined as a "form of observation in which a person considers his own subjective experiences." (4) The focus of introspection is still on *content*—the contents of inner experience rather than the contents referring to external events.

not only to perform phase shift easily and habitually, but also to *maintain it as an orientation* without losing contact with contents. When this is done, another new domain of experience becomes available.

It is convenient to give this new domain a name. We therefore introduce the term *main track* to denote the domain of experience occurring as a result of the orientation to contents that we ordinarily use, and *hypertrack* to denote the domain of experience occurring as a result of sustained orientation to operations.

As with any new skill, learning to orient to hypertrack is awkward at first. (Remember your first effort to ride a bicycle?) But gradually we learn to use it and soon become fascinated with the new perspective it gives us and the potentialities for development it affords. An immediate advantage is a greatly heightened ability to understand other human beings. *Operations produce contents.* Hypertrack orients us to the causal level of human thought, feeling and action.

There are other advantages that will become apparent as we proceed. One point soon emerges, however, Our language is adapted for use on the "mind band"—main track and orthoprocesses. It does not lend itself readily to communication about the domain of hypertrack (or the other domains of the broad band). There are many operations and processes of the broad band for which words do not yet exist. Hence, it has been necessary to introduce a number of new technical terms to describe these operations and processes. "Hypertrack" is an example of such a term. We refer to the evolving collection of such terms, affectionately, as "synergese." When syngeneers speak in synergese, it can be rather annoying to someone who is unfamiliar with the language. But every field has its technical jargon, including sports like baseball or football.

There is another sense in which language is inadequate. As noted, language is designed for the mind band of main track and orthoprocesses. It is not well-suited for managing the events of hypertrack or other domains of the broad band. Here, an analogy with computer science is helpful. The "language" that computers use is the language of numbers, actually a special kind of number composed of binary digits (zero and one). This is called machine language. It is very tedious and difficult to program a computer in machine language. Consequently, a number of special languages, called programming languages, have been invented. These are close enough to ordinary language (like English) that they are relatively easy to learn and to use.

Programs to control computers are written in one of these programming languages (such as FORTRAN, which stands for FORmula TRANslator). The computer then translates these programs into machine language automatically.

In the case of the broad band available to the human mind, we are confronted with a more difficult problem, the opposite of that which computer programmers had to solve. Computers were designed by humans, and the language they use, machine language, is known. We still know very little, however, about the broad band. Nevertheless, by trial and error, we are gradually developing a special language for controlling the broad band more effectively. It is called SYNTALK I. It is still not very well-defined, and a definitive version has not yet been presented. We won't do so here. However, portions of SYN-TALK I will be included in later sections of this book. This is a promising area for research by creative workers in synergetics—especially computer programmers.

One further remark about hypertrack: the ability to use hypertrack is basic to *tracking,* a powerful technique for controlling thought processes. This will be described later.

Phase shift, as we have noted, is the mental operation of focusing attention upon operations rather than contents. The inverse operation, moving from operations back to contents, is relatively easy to use. But there is another operation that is possible, a shift from main track to a more elemental and primordial domain. We refer to this operation as *prime shift,* and the domain "below" main track as *prime track.*

Prime track is the march of events as sensed before their organization into contents of the mind. It is the series of black marks on a white background from which you are now forming words with meaning. It is the set of processes actually going on when you are SICK and have MEASLES or a COLD. It is the "real world" out there and not the SIGHT or SOUND that gives you knowledge of it. It is your friend as he actually is, not the GOOD OLD PAL you think of him as. It is yourself in a strange place without your bearings, not the "I" that is somehow LOST.

In dealing with prime track, we sometimes adopt the convention of capitalizing all words describing what is perceived on main track. This permits the individual to perform a prime shift if he so desires.

Prime shift evokes the realization that, in ordinary consciousness, our attention is focused, not on the *actual* present, but on the *im-*

mediate past. By the time the raw data of sense have organized themselves into contents, time has already moved on. We are always one step behind in our perception of events. It is a very short step—a fraction of a second—but during that brief moment a variety of processes go on. This is another part of the domain of the fabulous microprocesses. In this fleeting moment many exciting and important things happen, of which we are ordinarily oblivious. Prime shift enables us to develop an awareness of these processes. We also learn that, once a content has been created, it tends to persist even when it no longer adequately represents what is currently happening. This is a major source of illusion.

One of the subtle fallacies to which the human mind is subject is the tendency to regard the sum total of its perceptions at any given moment as a complete representation of the world at that moment. When we reflect on this, we realize that this is not so; but the tendency persists anyhow. Several workers such as Arbib (5) and Fischer (6), have pointed out that perception is not just a passive process of high-fidelity mapping of the environment, but an active process of continuously constructing and reconstructing a map on the basis of sensory input cues, with selective emphasis on those referents that are relevant to the goals and interests of the individual. Furthermore, the perceptual systems on which the human mind depend for information endow its maps of reality with a particular *quality* that is by no means necessarily universal. An animal with a well-developed sense of smell, such as a dog, probably has a different quality for its maps; and one can conceive of organisms sensitive to ultraviolet or infrared radiation, or to ultrasonic sounds, or to magnetic fields or other forms of energy, also having a quality for their maps that might be quite different from those of human beings.

Prime shift enables the individual, to some extent at least, to free himself from exclusive linkage to main track contents. It brings to attention events and processes at the subgestalt level, processes that are filtered out by exclusive focus on main track. It also brings to awareness cognizance of *what is left out*—the realization, not just at an intellectual level, but at a concrete, action-influencing level, that *far more is going on* at a given moment than a person can possibly be conscious of.

Korzybski was fond of insisting "whatever you say a thing is, it is not." This paradoxical statement could be irritating, but its intent was to focus cognizance upon *what is left out* of any verbal representation no matter how precisely and thoroughly it is expressed. He also adopted the convention of frequent use of "etc." to remind the reader or listener of the necessarily partial and incomplete character of his statements. It is a wise convention.

Prime shift is also useful in breaking up *identifications*—the unconscious linkages (and blockages) of the Identic Mode. When combined with phase shift, it provides a powerful tool for clearing impedances, those "irrational" patterns of perception, thought, emotion, body control, and action that slow down and interfere with the effectiveness of mental function. This is discussed in more detail later.

Prime track, main track, and hypertrack thus comprise three levels of the broad band, just as microprocesses, orthoprocesses, and macroprocesses comprise three different characteristic tempos of events. It should be noted that hypertrack or prime track processes can also move at any of the three tempos. There are thus three times three, or nine, different "narrow bands" of the broad band as thus far described.

But this is not all. Effective function in the broad band requires the development of *synergies* among the various tracks and tempos. Thus, there is main track-hypertrack synergy, consisting of interactions that promote processes at both levels. Similarly, there is macroprocess-orthoprocess synergy. And so on. Etc.

As these synergies occur (as well as other synergies discussed later), there emerges a synergic whole that is greater than the mere sum of its parts. For lack of a better term, we sometimes refer to it as the "holistic" or whole being level. This emphasizes one of its aspects. But in another aspect, it is an old friend—the synergic mode.

A characteristic of the whole being level is that the individual no longer identifies with his consciousness and will. These become merely particular functions associated with main track and orthoprocesses—the "mind band" of experience. They are the command functions of the ordinary ego.

As a working hypothesis, we propose the view that the human mind is still evolving, and that the ordinary ego is a "transitional control center" which, like a butterfly emerging from a cocoon, will some day expand into a new control center, the Director, competent to the management of the broad band.

Be this as it may, there is another aspect of the whole being level that needs to be described, an aspect that is actually another domain of experience of the broad band. If, from the orientation of hyper-track, the individual performs another phase shift, an operation that concomitantly embraces all tracks and tempos, this new domain emerges into awareness. We refer to this domain as *ultratrack.*

It is extremely difficult, at first, to sustain an ultratrack orientation. What seems to be necessary is for a relatively high degree of synergy among the various tracks and tempos to be first established.

It is very difficult to describe in words the view from ultratrack or the processes that occur at this level. At this point, it seems wiser merely to define it as we have, as the level from which all tracks and tempos are viewed and managed, and to rely on the experience of the reader to fill in details.

Synergic Team Functions

A relatively stable orientation to the broad band is difficult to achieve at present. The operations of prime shift and phase shift are relatively easy to learn, however, as are the operations of selective focus upon microprocesses and macroprocesses. With practice and the use of exercises and techniques described later, one finds the broad band becoming increasingly accessible. Even here, however, this is best achieved in an environment in which the individual is relatively free from pressures and distractions. In ordinary social discourse and interactions, there are very strong constraints that force the individual to function almost entirely on the mind band. It is possible to resist these constraints to a certain degree, but the effort and struggle involved are considerable. It seems wiser, at first, simply to accept these constraints as "forces of nature" like the force of gravity, and to reserve efforts to expand into the broad band for synergetic sessions.

One of these constraints is the simple act of *verbal communication,* which plays so dominant a role in social discourse and interaction. This act, almost by definition, constrains the individual to or-thoprocesses and main track. And it is curiously difficult to be silent in the presence of another without feeling uncomfortable about it. The maxim that "silence is golden" seems to apply to another era.

Nevertheless, there are a few simple techniques that promote function on the broad band in the presence of another, which we have found useful. This is especially so when the other person knows something about synergetics and is interested in applying it. Indeed, use of these techniques while interacting with another syngeneer may quickly lead to a synergic relationship in which each helps the other to operate on the broad band. For this reason, these techniques are described here, although they properly belong in the field of Group Synergetics.

These techniques are not new—they have always been available to you, and no doubt you have used them on occasion. They do, however, promote synergy. And their habitual use, as part of a life-style, helps the individual function regularly in the synergic mode, using the broad band.

1. Affinity make. In the course of relations with another human being, aspects of his action or being periodically emerge for which one feels affinity. This is true of most of the people one encounters. Each of us has so many facets that some are bound to be "likeable."

When such affinity is felt, express it. This action is an affinity make. When it is done, both parties feel better and a surge of synergy occurs.

Two important qualifications should be noted:

a. An affinity make is primarily expressed by *action,* not words. It can be by a look or a gesture. Of course, a verbal statement is a form of action and is often the simplest way to do it, but it is more the way it is said than the explicit content that makes the flow of affinity. An example: "I hate you," said in an affectionately jocular manner.

b. It must be genuine. Most people are aware of the power of flattery and are pretty good at detecting it. Whether detected or not, flattery does not promote synergy. This is not stated as an ethical judgment, but as an observation of human beings in action.

Opportunities for affinity makes are constantly occurring. But there is so much dysergy in the world that these opportunities are often overlooked. Yet an affinity make is a powerful synergy generator.

2. Empathy make. This consists essentially of the operation: "Put yourself in the other fellow's shoes." This does not mean doing so from your viewpoint and values, but from his viewpoint and values. It is not necessary to *accept* his viewpoint and values, merely to *understand* them and to see how events and situations look from his perspective.

An empathy make has several values:

a. Each human being is like a walking, talking library, with years of experience, data, and know-how different from yours. It is always possible to learn from another human being, no matter how humble, no matter how great. An empathy make enables one to take advantage of this opportunity.
b. An empathy make promotes affinity, mutual understanding, and effectiveness of communication.
c. An empathy make develops the ability to see things from a variety of perspectives simultaneously. It turns on the multi-ordinal mode. From this, it is a small additional step to the synergic mode.

3. Semantic telepathy. Affinity and empathy makes help one to communicate with others in a synergic team with remarkable effectiveness. We refer to such communications as "semantic telepathy." (The word "telepathy" is used here not in the usual sense of "direct thought transference" but in the sense of nonverbal communication of meaning.)

Consider two individuals, Mr. A and Ms. B. Mr. A has an idea that he wishes to communicate. He first of all encodes the idea into words and speaks the words. Ms. B hears the words and decodes the message. If all goes well, the idea she gets will be the same as Mr. A's. When this happens, we say that *semantic communion* has been achieved. Semantic communion does not necessarily mean agreement, merely understanding.

Semantic communion is the primary goal of communication, and one can use a variety of means other than the verbal message to achieve it. The set of these other means constitutes semantic telepathy. The receiver, for example, may make a deliberate effort to predict the message. One way to do this is to follow the rule: "Focus on what he means, not what he says." As soon as she gets the message, she calls out "clear," and Mr. A immediately ends the verbal message.

The sender, in framing the verbal message, uses empathy makes in order to state the idea in terms that fit the perspective of the receiver. He is continuously aware that the same word may have different meanings to different people or even to the same person at different times. Since semantic communion is the goal, he does not insist on the "correctness" of *his* meaning, but accepts hers.

If the idea is abstract, such differences of word meaning may be considerable. So he follows the natural movement of the mind, in which a concrete perceptual experience usually precedes an abstraction from that experience, and begins with a concrete presentation that readily evokes semantic communion, and *then* moves to the abstraction, rather than first stating the abstraction.

One very useful technique is called *bridging*. The sender evaluates areas of agreement he has with the receiver, and separates these from areas of difference of outlook, disagreement, or conflict. He then uses the area of agreement as a bridge through which to transmit his message. A good rule here is: "Pick an agreement with her."

A frequent obstacle to semantic communion is the existence of a distinction made by the sender but not by the receiver (or vice versa). This is a source of confusion. The idea of empathy, for example, may imply or include the concept of sympathy to the receiver, whereas for the sender these are two somewhat similar but distinct ideas. Whoever makes the distinction is best able to communicate it.

Another obstacle is an apparent agreement that obscures the fact that semantic communion has not really been achieved. The receiver may nod agreement because the verbal message evokes a clear picture in her mind, not realizing that the picture is not the same as that of the sender. This can be minimized by a policy of not taking semantic communion for granted, a policy adopted by both sender and receiver. Semantic communion can be checked by the sender using a concrete illustration of the message he has sent, or by the receiver repeating the message in her own words. The mere cognizance of the possibility of this source of confusion minimizes the probability of its occurring.

There are other purposes of communication besides semantic communion such as achieving agreement, persuading the receiver to accept an idea or to do something, or simply to convey affinity (or rejection or some other state of relationship). But, for most of these, semantic communion is prerequisite. It is indeed remarkable that despite the tremendous expansion of the physical means for communication—telephone, mimeograph or other forms of replication, radio and TV, etc.—semantic communion is so often *not* achieved. Misunderstanding piles upon misunderstanding, and affinity and empathy go down, with a concomitant rise in mistrust, hostility, and conflict. While such failure to achieve semantic communion is by no means the only cause of human problems, it is a major cause of many and a contributing cause of most.

4. Synapse. Affinity makes and empathy makes can be used with anyone. So can semantic telepathy, but it is much more effective when done by two syngeneers. When each party knowingly focuses on semantic communion as the goal of communication, the interchange of information and the degree of trust and rapport can reach remarkable heights. And as this occurs, a step beyond semantic telepathy becomes feasible.

Any message tells far more than it says. Surrounding the *explicit statement*—what the message says—there is a network of plausible inferences and connotations, the *implicative residue.** When semantic communion is rapidly and easily achieved, communication can be expanded by focusing on the implicative residue.

The basic rule of semantic telepathy is: "Focus on what he *means,* not what he says."

The basic rule of synapse is: "Focus on the implications of what he *means."* It is a step beyond and much fun.

5. Franktalk refers to presentation without rancor of ideas or evaluations critical of the actions or viewpoints of another. An implicit convention governs franktalk. If this convention is broken, franktalk is ineffective and often counterproductive. It is therefore recommended that franktalk not be used unless one is sure that the convention holds. The convention is usually easy to establish by use of affinity makes or bridging (or both) beforehand.

The convention is simply neither to take offense nor to give it. If the sender "talks down" to the receiver, displays or feels hostility, shows an unwillingness to receive franktalk in return, etc., the convention is broken. If the receiver feels hurt, imputes unfriendly motives, or feels called upon to defend or justify, the convention is broken. The sender does *not* try to persuade; he simply *presents for consideration.* Similarly, the receiver *accepts for consideration.* That is all.

Franktalk gets behind the veneer of politeness we so often use to hide from one another. Among syngeneers, it can be highly effective and useful.

6. Totaltalk. This is an advanced mode of communication that emerges when the previously described synergic team functions are

---

*Show a person a nail. This implies (a) a technology for extracting metals from ore, (b) techniques for manufacturing nails, (c) the existence of hammers, (d) the existence of wood or other material, (e) ETC.

used so regularly that they form a synergic whole and when a broad band orientation has become characteristic.

We can distinguish four channels:

a. *Mind-to-mind.*
b. *Mind-to-whole.*
c. *Whole-to-mind.*
d. *Whole-to-whole.*

In totaltalk, all four channels are used concomitantly. For example, one reads the mind band and uses it. Simultaneously, one reads the implicative residue, as much as one wishes. This can be done as a mind, consciously. It can also be done as a whole being, "knowingly." By "knowingly" is meant the whole being analog of consciousness. But one should not be bound by this analogy. To "know" in this sense is "to-be-able-to-be-conscious-of-if-the-need-arises." It is this, and more, but words fail. Get the feel?

It is possible to describe in greater detail the enormous variety of processes that go on in totaltalk. But a verbal description would take a whole book in itself and would still be inadequate. Instead, let us merely regard the four available channels, focus on the implicative residue, and let out minds go where they will.

One final word: ETC.

Self-actualizing people are, without one single exception, involved in a cause outside their own skin, in something outside of themselves. They are devoted, working at something, something which is very precious to them— some calling or vocation. . . They are working at something which fate has called them to somehow, and which they work at and which they love, so that the work-joy dichotomy in them disappears. . . Self-actualization means experiencing fully, vividly, selflessly, with full concentration and total absorption. . .

———Abraham Maslow (1)

CHAPTER 6

THE SYNERGETIC SESSION

There are many, many schools of self-improvement, of psychotherapy, of seeking enlightenment, of meditation, of mental discipline, etc. Many of these report a "turning on" of what are coming to be known as *altered states of consciousness* (ASCs). These include states of ecstacy, of mystical union, of transpersonal knowledge, etc. An ASC has been defined by Tart (2) as follows: "An altered state of consciousness for a given individual is one in which he clearly feels a *qualitative* shift in his pattern of mental functioning, that is, he feels not just a quantitative shift (more or less alert, more or less visual imagery, sharper or duller, etc.) but also that some quality or qualities of his mental processes are *different."* This is admittedly a subjective definition, and therefore anathema to some schools of psychology. But these phenomena do occur, they are a part of the human experience, and they deserve to be treated fairly and open-mindedly on their own merits. A scientific paradigm that refuses even to consider their existence is obviously incompetent to analyze them.

When a person has effectively oriented to the broad band, when the synergic mode turns on, he also experiences an altered state of consciousness. It is distinct from, but in some ways similar to, the other ASCs that have been described. In particular, there is an ineffable quality about it—so much goes on so fast that it is virtually impossible to describe it adequately in words. Moreover, there is a *quality* to it that a verbal description cannot capture. The synergic mode has to be experienced to be understood.

In the stress and hurry of everyday life, buffeted by a thousand pressures from a thousand sources, it is very difficult to be fully synergic. The very fact that so much of our lives involves *speech,* talking and listening to others, tends to focus attention almost entirely on the mind band. This is not to say that synergic function cannot be achieved, that it never happens. But it is comparatively rare.

This means that a *special environment,* and special techniques, are usually required to evoke the synergic mode. In this, synergetics is not unlike other schools. In psychoanalysis, for example, the patient lies on a couch and reports without censorship whatever comes into his mind as he free associates. In yoga and similar schools, an effort is made to still the mind. In hypnotism, the subject is placed in a state of hypnosis. Other schools use drugs to evoke their particular ASCs. In synergetics, this is accomplished in the synergetic *session* by means of a technique called *tracking.*

The basic requirement of the special environment for a synergetic session is that the individual be in a place where he is free of distractions and interruptions, able to concentrate on applying synergetics to his own case. He should either be alone or with at most one or two others who act as monitors or coaches (see chapter 11). As long as this basic requirement of freedom from distractions and interruptions is met, it doesn't matter whether he sits or lies down or walks around—whatever comes natural to him, he freely does.

The basic technique of a synergetic session is called *tracking,* and the individual undertaking the session is called the *tracker.* Tracking is described in the next chapter. There are, however, certain conditions that experience has shown must be met if tracking is to be effective; these are discussed below. In addition, there is a simple technique called a *sweep* that is useful as an exercise to facilitate tracking.

The conditions necessary for effective tracking are as follows:

1. The tracker should have a clear understanding of the broad band in concrete terms of his own experience. He should be able to recognize microprocesses when they occur and to examine them in retrospect at a slower tempo. He should be able to detect

macroprocesses by virtue of *changes* that have slowly occurred and to examine the processes that brought about those changes. He should be able to distinguish clearly between contents and operations, and to perform phase shift to focus attention on operations. (This also means being able to distinguish phase shift from other mental operations such as introspection and abstraction.) He should be able to probe behind the ordinary contents to the subgestalt level from which they emerge—in short, to do a prime shift on any content. And he should be able to move to ultratrack, viewing all tracks and tempos concomitantly.

(Actually, the tracker soon realizes that when he performs phase shift or prime shift or looks for micro or macroprocesses, he already is at ultratrack.)

When the tracker can do these things, and do them easily and well, he will have achieved a broad band orientation. At the same time, his consciousness will have expanded, since he no longer is restricted to the mind band. Experience has shown that a good broad band orientation is essential to effective tracking. The reasons for this will become clear later.

2. The tracker should have an open mind, ready to re-examine and re-evaluate any belief, attitude, motive, or value, etc., no matter how strongly it is held. This does not mean a commitment to change, merely a willingness to look afresh and a readiness to change if he decides it is warranted. It does mean he feels no need to defend a belief or to rationalize a motive. And it means, above all, that he regards such a re-examination and re-evaluation as an opportunity to learn, an opportunity to gain new insights, an opportunity to acquire a new and better perspective. In short, he regards the experience he is about to undergo as an Information Source.

3. The tracker must have a high value for the truth, no matter where it leads or what it costs. He must be willing to search unpleasant aspects of himself, no matter how painful. He must be ready to give up the superficial pleasures of wishful thinking. On the other hand, he must also refuse to be dominated by fearful thinking. In short, what is required is total, objective self-honesty. Without this, tracking will not get very far. With it, tracking can achieve a great deal.

These three things, then—a broad band orientation, a flexibly open mind, and total, objective self-honesty—are essential to good tracking. Of these, the latter two are not peculiar to synergetics, of course; and they can only be provided by the individual himself. To help with the first, a technique called a Broad Band Sweep is described below.

A *sweep* is a mental operation everyone has performed, but which is sometimes surprisingly difficult to complete. It consists simply of a systematic examination in sequence of a set of contents and/or operations. Thus, for example, one might think in turn of each of his grandparents, trying to form a concept or a mental image of each. Or one might sweep through the main events of the day while lying in bed before going to sleep, trying to recall essential details.

A major difficulty in sweeping is that the mind often becomes distracted by some idea, recollection, feeling, etc. and goes off on a tangent, sometimes forgetting to complete the sweep. There are two extremes to be avoided here. One is a too-rigid insistence on completing the sweep, suppressing all distractions—the steel-trap mind. This is unwise because it may lead a person to overlook an important train of thought which, if followed, could lead to unexpected and exciting results. It keeps the individual in the Uniordinal Mode, confined to a single perspective.

The second is an excessive tendency to be distracted, following ideas and feelings randomly. The individual starts a sweep but never completes it and soon forgets what he started to do. Now, of course, such an attitude is encouraged in the free-association technique of psychoanalysis, and there are times when it can be useful. I am not denying its value in certain circumstances; but when one decides to sweep, one decides to sweep.

The optimum policy in sweeping is to follow distractions whenever it seems appropriate, to a certain extent, but always to return to complete the sweep (unless one deliberately decides otherwise for a good reason). Thus, in sweeping the events of the day, one may recall an unresolved problem and spend some time working on it further, but at some point one decides to return and complete the sweep.

There are two sweeps in particular that I recommend for consideration. The first, as mentioned above, is the Broad Band Sweep. Start with a particular content or set of contents. This may, for example, be what is happening at the present moment; it may be a person you know; it may be an idea you have or a problem you are considering or a potential you would like to explore. Or it may be a verbal statement. Then do, in order, the following operations:

1. A phase shift
2. A prime shift
3. A look for microprocesses
4. A look for macroprocesses
5. A look at all tracks and tempos concomitantly

To illustrate a Broad Band Sweep, let us consider the following statement: "The Id is the original system of the personality."

This statement is a content generator. Indeed, verbal expressions are very useful content generators for a Broad Band Sweep, in part because they immediately focus attention on the mind band (main track and orthoprocesses). The actual contents produced will vary somewhat from mind to mind, so that what follows should not be regarded as universally valid but simply as the function of one particular mind (the author's).

The contents generated are a set of ideas denoted by the words of the statement. In addition, a number of associations evoke other contents, e.g., recognition that this is a Freudian theory, a mild sexual fantasy, the thought of Jung's theory of personality, and a number of extraneous contents.

Phase shift leads to the detection of an operation—the operation of equating *Id* to *system,* plus other operations attaching qualifiers ("original" and "of the personality") to the idea of system.* Moreover, there are other operations† that are *not* directly evoked by or associated with the verbal statement, but which occur and are important. The verbal statement is not only a "proposition" (denoting the operation of *equating* "Id" and "system"); it also implies the operation of affirmation—asserting the statement is true. In addition, there is evoked an operation (in the author's mind) of *tentative acceptance* of the truth of the statement. (Other operations also occurred, but for our purposes the foregoing are sufficient.)

The foregoing demonstrates that it is not easy to describe in words what occurs on hypertrack—we do not ordinarily focus attention on this domain and lack words to describe hypertrack events and processes. It is also apparent that a lot more goes on in our minds than we are ordinarily aware of.

---

*It should be noted that we are *not* undertaking a linguistic analysis here, such as that of parsing the statement. Our focus is on the *contents* evoked by the verbal statement and on the *operations* producing those contents.

†These are related to, but distinct from, the "deep structure" postulated by Chomsky and other workers in modern linguistic theory.

Prime shift. Look at the verbal statement again: "The Id is the original system of the personality." But look at it simply as a pattern of black marks on a white background. Look even below this at the processes that organize the black marks into distinct patterns. It's a bit like viewing a picture with eyes focused on the distance, but instead of the letters, it is the meaning that blurs. THE ID IS THE ORIGINAL SYSTEM OF THE PERSONALITY. One eventually may capture the impression of a statement in a foreign language whose words are unfamiliar. Prime track emerges—and we appreciate that main track, while not exactly an illusion, has reality because we endow it with reality—it is we who make it real.

This is the famous Maya of Indian philosophy. The visual world that we see and accept as real ("seeing is believing") is "actually" like a three-dimensional color TV moving picture. Indeed, watching TV, we endow the patterns we see on the tube with reality in much the same way.

But prime shift does not mean rejecting main track as total illusion. It merely expands awareness to include both main track and prime track (and a lot of other tracks and processes that we have ignored for the sake of simplicity).

Microprocesses. "The Id is the original system of the personality." In each fraction of a second, as one reads this statement, a large number of microprocesses occur that endow the statement with meaning. These are difficult to become conscious of, but occasionally they *flash* into awareness. In a Broad Band Sweep, one need not expect such flashes to occur—simply look for them as the contents stream by. As experience with Broad Band Sweeps grows, such flashes tend to occur more and more often. There are an *incredible* number of things going on in our minds of which we are unaware!

Macroprocesses. "The Id is the original system of the personality." Repeated reading of this statement has by now evoked a fairly stable state of mind in connection with it. The first time it was encountered is quite different from the nth time. Slow processes have occurred between the first time and the present one, which brought about this change. Ordinarily we are unaware of these processes, and we have few words to describe them. Again, in a Broad Band Sweep, do not expect other than a vague awareness at first—with practice they become clearer.

Ultratrack. "The Id is the original system of the personality."

Looking at all tracks and tempos:

TEMPOS	TRACKS
Microprocesses	Prime track
Orthoprocesses	Main track
Macroprocesses	Hypertrack

ETC.

At first, this is simply a matter of putting it all together as an interrelated whole. Later, there may emerge a feeling almost of a superconsciousness—thoughts moving along several tracks at once, contents at a very general level, each capable of being expanded into a long train at main track ortho tempo but accompanied by realization that this is only occasionally desirable and all sorts of cross-correlations are going on words cannot readily describe....

Broad Band Sweep is an exercise, and nothing dramatic is to be expected (although it sometimes occurs). It does have a value in that it evokes a broad band orientation. What you do with the orientation is something else again. But just as an athlete or a musician develops skill by practice, so does one develop skill in tracking by practice; and a Broad Band Sweep is a simple exercise that helps one to acquire this skill.

The second sweep is the Mode Sweep. Consider the Mode Ladder:

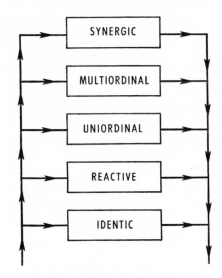

THE MODE LADDER

Again, start with a content or set of contents, such as those of the present moment. Then move step by step up the Mode Ladder.

Identic Mode—Recognize that it may be and probably is operating in you without your awareness. Search for *identifications*—associations* you may be making without discrimination or awareness. A particularly good place to look is at the associations that occur immediately after a perception, external or internal. If you hear footsteps, what do you think of, feel, recall, etc., immediately after the sound? If you recall your mother, what do you think of, feel, imagine, etc., immediately after the recollection? Often this association is made without discrimination or awareness, and the connection may by no means be obvious. It isn't necessary to explore this in great detail, just *become aware* of the identification, and make a *distinction* between the contents. It is surprising how often we are confused because of a simple distinction we have failed to make.

Reactive Mode—Search for emotional reactions to people, situations, problems, events, etc. that may be operating at the moment. Here, the effort is *not* to deny the aim of the emotion (which is a response to a felt need), but to *align* with it and then to seek a more intelligent, a more synergic way of achieving that aim. In other words, introduce rational guidance of the emotional reaction.†

Uniordinal Mode—Look at the viewpoint you are currently holding with respect to the given content or set of contents. How does this viewpoint influence your interpretation and evaluation of this content? What other viewpoint might you conceivably adopt with or without abandoning your current viewpoint? How would this affect your interpretation and evaluation of the content?

Multiordinal Mode—Look at the interactions among the two viewpoints. Are there any ways in which these interactions mutually aid or reinforce each other? What other viewpoints might you tentatively adopt? Look at the interactions of this new viewpoint with the older ones.

Synergic Mode—Look for any emergents (qualitatively new features) that may have been evoked by the synergic interactions previously encountered. What opportunities do these emergents provide? Look for a basic perspective that integrates all previously encountered viewpoints.

---

*These are called *linkages*. Negative associations, separating two contents without discrimination or awareness, may also occur; these are called *blockages*. They are, obviously, more difficult to detect, and need not be looked for in a Mode Sweep.

†Frequently, when this is done, the reaction disappears and the aim no longer seems important.

A Mode Sweep does not have to include all the features just described for each step. Conversely, you may wish to add other features not mentioned. Each person may do a Mode Sweep a little bit differently, and each Mode Sweep done by a given individual may be different. Adapt the tool to your needs and preferences, not vice versa.

The value of doing sweeps is twofold. First, a sweep is a good way to learn different parts of synergetics. Thus, exercises with a Mode Sweep fix the Mode Ladder in one's mind in a ready-to-use state. Second, and more important, sweep exercises help to develop the individual's ability and skill in self-programmed thinking. This mobilizes the powers of the mind, increases the effectiveness of thinking, and evokes phenomena that otherwise would not have occurred.

It is recommended that before undertaking a synergetic session, at least one Broad Band Sweep and one Mode Sweep be done. For your convenience, the steps of each are briefly repeated below. Practice with these sweeps will increase the effectiveness of tracking.

Broad Band Sweep
1. Select content or set of contents
2. Phase shift
3. Prime shift
4. Look for microprocesses
5. Look for macroprocesses
6. Look concomitantly at all tracks and tempos

Mode Sweep
1. Select content or set of contents
2. *Identic*—Look for linkages, especially following perceptions. Make distinctions.
3. *Reactive*—Look for emotional reactions. Align with the aim of the reaction. Think of a better way to achieve the aim.
4. *Uniordinal*—Look at your viewpoint and how it affects interpretations and evaluations. Think of another viewpoint.
5. *Multiordinal*—Look for synergy in the interactions of the two viewpoints. Think of another viewpoint and look for synergy.
6. *Synergic*—Look for emergents. Think of a basic perspective that integrates viewpoints.

A Synergetic Session, then, is done in a place where the tracker is free of distractions and interruptions, and can focus on synergetics.

The effectiveness of a session depends on a clear understanding of the broad band, a flexibly open mind, and total objective self-honesty. A sesson can be of any length, but usually lasts fifteen minutes to an hour or two. A notebook is very useful for writing down insights, ideas, things to do or track on, etc. Sometimes a tape recorder is helpful to record verbal descriptions of the events of a session for later review.

*Most of us don't think; we simply
re-arrange our prejudices.*

———Adlai Stevenson

CHAPTER 7

TRACKING

Suppose a visitor from outer space were to drop in on you and leave you a mysterious and wonderful electronic gadget that could solve any problem in the universe. All you would have to do is plug it in and state the problem, and the gadget would beep and hum and purr and maybe click a time or two, and then chimes would ring and out would pop a card with the answer printed on it.

It would be rather nice to have such a gadget, wouldn't it! You could ask it how to get more money, or how to become more attractive to members of the opposite sex, or how to eliminate those clashes with husband or wife, or what to do about Junior's temper tantrums or Helen's difficulties with her teacher, etc. You could even ask the gadget how to eliminate war or poverty or greed or any of the other major problems that afflict humankind.

Well, there is no visitor from space, of course. But you do have such a gadget. *It's right inside your head!*

The human mind has often been compared to an electronic computer, and many hot arguments have raged over this comparison. Questions like: Do computers think? Are computers intelligent? Can computers feel? etc., usually generate more heat than light, and are really unanswerable as stated because they are phrased in inadequately defined terms. Our concern here is not with such esoteric questions but rather with whether this analogy can be useful. Up to a point, I have found it to be very useful, provided that the analogy is not carried too far. But if what follows disturbs or annoys anyone, please remember that it is only an analogy. Human minds designed computers, human minds program them, and everything a computer does was first done by a human mind.

Indeed, the human mind is far, far superior to any computer yet designed. By comparison, a computer is nothing but a high-speed idiot. Nevertheless, because computers are like minds and indeed may be regarded as extensions of the human intellect, knowledge of how computers work can be useful in improving mental function.*

How is a computer controlled? It is controlled by *programs*—lists of instructions that tell the computer what to do. A program telling a computer-controlled robot to open a door, for example, might consist of the following instructions:

1. Move to within one arm's-length of door and stop.
2. Extend arm toward knob.
3. Grasp knob.
4. Turn knob clockwise 1/8 turn.
5. Pull door toward you.

The only trouble with this program is the last instruction. The robot would pull the door off its hinges, because there is no instruction to stop pulling! A computer always does *exactly* what a program tells it to do; it has no common sense or judgment. Its behavior, in short, is not unlike that of a human being operating in the Identic Mode.

In the case of the human mind, we are dealing with a self-determined system; efforts to program it *from the outside* are not only a violation of human freedom but also, sooner or later, backfire; the individual overtly or covertly resists. However, the human mind *can* be programmed from the inside, by the individual himself. This does not violate his freedom; on the other hand, it introduces order and efficiency into mental processes.

Tracking may be simply defined as self-programmed thinking. The tracker carries out, step-by-step, a short sequence of mental operations. The performance of these operations starts and continues a train of mental processes, and achieves a small but definite activation of the rational mind.

But this is synergetics. Just any sequence of operations won't do. We are interested only in those operational sequences that tend to evoke synergy and/or reduce dysergy. Hence, tracking is carefully limited to mental programs designed to accomplish this.

---

*Computer programmers would add that the mental effort required to program a computer disciplines the mind in a valuable way. It forces one to think clearly!

The CEDA Sequence

The basic technique of tracking is the CEDA Sequence (pronounced "say-duh sequence"). This is a sequence of four mental operations—C for Consider, E for Evaluate, D for Decide, and A for Act. Tracking is continued by one CEDA Sequence after another.

Consider means to focus your attention upon some field of interest. It may be a problem you want to solve, a belief you have adopted, a potential you would like to develop, a person whose behavior you want to understand, an incident you want to examine, a bad habit you'd like to eliminate, an idea you don't fully comprehend, etc.

There is, in the Consider phase, an attitude of search, of free and open-minded inquiry. You are seeking new data, new ideas, new viewpoints. You are ready for change, not necessarily committed to it if the new is not as good as the old, but willing to look. This attitude is of basic importance; if it is not present, tracking is ineffective.

The Consider phase can last a few seconds or thirty minutes or more. You are free to explore in any direction you choose, to think in any way you please. But there is one essential thing to be done. At some stage *you formulate two or more possible lines of action.* This is necessary for the next phase of the CEDA Sequence.

Evaluate means to think about these possible lines of action. Again, you are free to do this any way you choose. You may want to clarify or change one of these lines of action, or even to formulate another line of action if it occurs to you. You may want to seek additional data on some point.

You think about the value to you of each line of action. You also consider the value it may have to others. Sometimes, after a period of evaluation, one line of action may seem clearly superior to the other. At other times they may be about equal. Not infrequently you may find that the two lines of action are related to each other in some way. Evaluation may be very extensive or very brief. At some stage, however—and this is often arbitrary—you move on.

Decide means to select one of these lines of action.

Act means to carry out this decision.

The line of action may be "doing something." Or it may be following some line of thought. In any event, it leads naturally into another CEDA Sequence. Tracking is basically just one CEDA Sequence after another.

You can make a CEDA Sequence as fast or as slow as you like. You can make it cover as much or as little as you choose. You can make your actions as tentative or as positive as you choose. You can skip back and forth between phases if, as sometimes happens, this seems like the thing to do. You select or reject whatever beliefs you choose, whatever attitudes you choose, whatever motives you choose. Tracking imposes no particular viewpoint or belief upon you, no special way of thinking or feeling or acting. All that tracking asks of you is that you do choose—that you *keep tracking.*

Let's take an example of a CEDA Sequence to illustrate what is meant.

Consider. "Let's see now, what shall I track on? Possibly this funny feeling I get in my stomach when I am anxious about something. Or perhaps about the difficulty I have had lately in sleeping at night." (Think about this for awhile.) "Is there anything else? Yes, there is. I am also interested in getting along well with people. I wonder, is there any relation between this and that funny feeling in my stomach? Can't think of any just now."

Evaluate. "The ability to get along well with people is certainly one I would like to have. Getting plenty of sleep at night is also desirable. Perhaps the two are related. I have the feeling that they are. In any event, I can think of more things about the first than the second."

Decide. "Track on getting along with people."

Act. "Track on that."

One CEDA leads naturally to another. The important thing is to formulate at least two lines of action, to evaluate them for awhile, and then to decide and act.

The astute reader will have noted two things:

1. A CEDA Sequence *automatically* moves the tracker into the Multiordinal Mode, by leading him to consider at least two possible lines of action. By virtue of the attitude of open-minded, ready-to-change search in the Consider phase, it also orients him to the emergence of synergy.
2. Facility in doing a phase shift, in orienting to hypertrack as well as main track, greatly helps the tracker to stay on track.

It may help in tracking to keep a notebook in which to write down important ideas and insights as they occur. Indeed, this is so useful that it is highly recommended.

It may also help, especially at first, to write down:

> Consider
> Evaluate
> Decide
> Act

on a 3 x 5 card, and keep it before you. This way, if you go off track, you can readily get back on track again.

Tracking is basically simple. Indeed, everyone has probably gone through a CEDA Sequence at one time or another. But few of us do it deliberately and knowingly. Tracking *organizes* thinking and keeps it moving according to a rational pattern. Moreover, it turns the mind *on*. Indeed, when a person learns to track, he soon finds that his thinking becomes a lot clearer and faster and more accurate. Ideas come thick and fast; new insights are gained; new ways of looking at things occur; and a lot goes on nonverbally, along several lines at once. The mind surges into overdrive.

It's a delightful feeling.

The BAM Triangle

Tracking can be effectively done by simply executing one CEDA Sequence after another. And it is better, at first, to limit yourself to this process until you get the hang of it. But the CEDA Sequence is far from the only synergetic mental program that is available. Indeed, one of the delights of tracking is the discovery or design of other mental programs; each tracker develops his own "program library." In what follows, a few such programs that have been found to be of value will be described.

The CEDA Sequence remains the *basic* technique of tracking, however. Other programs are generally (but not always) *introduced* during the Consider phase. They are executed, and the CEDA is resumed by formulating two or more lines of action, Evaluating, Deciding, and Acting.

One very useful program is the BAM Triangle (B for Belief, A for Attitude, M for Motive).

The terms "belief," "attitude," and "motive" have a number of different meanings in common usage. I hope the reader will be tolerant if the definitions given here do not correspond exactly with his own. Actually, I have observed that the BAM Triangle is effective even when used in rather different ways by different trackers! As with all synergetic techniques, adapt the tool to your own needs and potentials, not vice versa.

By *Belief* is meant a representation of reality that is accepted by the individual as "true" or "valid" or "a basis for action."

By *Attitude* is meant the way a person orients to a situation.

By *Motive* is meant a desire to do something—a felt need for action.

A *Belief* consists of two parts: data, and interpretations of that data. A person may have a belief, for example, that people are naturally selfish." This belief can be analyzed into data and interpretations. The data would consist of particular people in particular situations acting in a particular way. The interpretation consists of an evaluation, a judgment, of the behavior of these people, combined with a generalization to the effect that all people naturally behave that way at all times.

Many people, perhaps most people, once they adopt a belief, tend thereafter to hold it more or less rigidly and are relatively unwilling to re-examine it open-mindedly. If someone else disagrees with the belief, they tend to defend it and to consider any change in the belief as a sign of weakness. Henshaw Ward coined an interesting word to describe the thought processes of such a person: he called it "thobbery." The word is formed by taking the initial letters of *th*oughts about *o*pinions that were already *be*lieved before reasoning began. "Thobbery is the confident reasoning of a person who is not curious about verifying his results."* (1)

Tracking is basically different from thobbery; as already noted, an essential feature of tracking is that the tracker have an open-minded, flexible, ready-to-change attitude. The tracker examines the belief with this attitude; he looks carefully at the *data* upon which the belief is based and then at the interpretations he has made of those data. More often than not, he finds that the data are insufficient to support the belief and/or that other interpretations of the data are at least as good, and perhaps better. Tracking does not require a person to change a belief; but frequently he decides to do so.

But beliefs do not stand alone as mental events. Usually they are associated with *motives*—desires to do something (or to avoid doing something). Thus the belief that "people are naturally selfish" may have been adopted as a reaction to an incident in which someone else

---

*We all know people who thob. Indeed, is there anyone who hasn't thobbed?

did something that prevented the tracker from getting what he wanted. If a belief is strongly held, a motive is usually involved.

In tracking on motives, it is useful to trace a motive down to its basic source. Having done this, the tracker then replaces the reactive motive with a synergic response. Motives can almost always be tracked down to one or another of the following basic motivations:

1. A threat to, or a loss of, self-esteem. Whenever a person's self-esteem is reduced or threatened, he tends to react by an effort to restore or protect it. Each of us likes to think well of himself; and if something happens to prevent this, we don't like it, and more often than not our response is reactive in mode.

When the tracker traces a motive back to a Self-Esteem Threat Point Reaction (SET Point Reaction) or a Self-Esteem Loss Point Reaction (SEL Point Reaction), he realizes how irrational this is, and either drops the reaction entirely, or he substitutes a synergic response. This rational re-evaluation clears the reaction. Often this is accompanied by laughter.

Indeed, a relation can be observed between the mode of an individual and his state of self-esteem. This is called the Self-Esteem Curve. In tabular form it looks like the following:

Self-Esteem	*Mode of Function*
Self-pity	Identic
Self-punishment, blame, guilt, inferiority feelings, sympathy seeking	Reactive
Moderate self-confidence	Uniordinal
Optimum self-confidence	Multiordinal to Synergic
Overconfidence; cockiness	Uniordinal
Egotism; conceit	Reactive
Megalomania	Identic

The self-esteem level of a person is not constant. It varies from time to time, from situation to situation. When mode is high, a person fluctuates around optimum. Sometimes he may be overconfident, even a bit cocky. At other times, he may be underconfident, a bit unsure of himself. At optimum, his self-appraisal is accurate; he knows what he can do and does it. He is unconcerned about himself, absorbed in what he is doing.

But when he encounters a SET or SEL point, he may react strongly. He may try to excite sympathy or attention from others. On the other hand, he may withdraw into a shell, as if to hide. He may boast about his prowess. He may blame some other person or event. He may even become angry and attack (verbally or even physically) someone else. Or he may turn on himself and punish himself, either depriving himself of something or turning on feelings of guilt, remorse, or shame.

The effort to restore or protect self-esteem, in response to a SEL or SET point, is one of the most powerful motives of human beings. Rather than honestly acknowledging to himself that he has been wrong about something, a person will go to extreme lengths to prove he really was right, or at least was justified. Indeed, there seems to be a kind of club about this. Members of this club (at present the majority of the human race) follow the rule: never admit you are wrong about anything. If you do, you get kicked out of the club. You have become a SET point to all the other members.

Yet we all know in our hearts that such self-deception is ridiculous. In the first place, it doesn't work. A part of us always knows, so we shut off awareness of that part. In this way we build up a network of SEL and SET point reactions, shutting down large parts of our minds.*

Such reactions are often the basic source of motives that hold an irrational belief in place. Clearing the basic SET or SEL point reaction by rational re-evaluation releases the irrational belief, enables us to replace it by a synergic belief, and opens up a shut-down part of the mind.

2. A threat to, or a loss of, general esteem. Man is a social animal. Every person has a basic esteem for other persons he knows, as well as for groups, ideas, etc. Conversely, every individual receives a certain level of esteem from other people—either for himself, or for his status in a group, or for the roles he is performing in the group, etc.

Let us refer to all these relations of esteem as general esteem. Now general esteem may be positive, neutral, or negative; and it may be high or low in intensity. Of particular interest is the fact that there occur General Esteem Threat points (GET points) and General Esteem Loss points (GEL points). When this happens, the individual

---

*As discussed later, this is not the fault of the individual. The "never admit you are wrong" club did not happen by accident; and its existence compels us to reactively restore or maintain self-esteem. If we don't, we are placed at a disadvantage. But more on this later.

TRACKING   81

may *react*, just as he reacts to SET and SEL points.

We can construct a general esteem curve that is like the self-esteem curve. The general esteem curve may be applied to each "esteem relationship" of the individual and other persons, groups, etc. At any given instant there exists a mutual esteem relationship between the individual and each person he knows in the group. Thus, there is your esteem for Tom and his esteem for you.

If your esteem for Tom goes up, and you let him know this, his esteem for you will tend to rise. If Tom's esteem for Jack falls, and you observe this, your esteem for both may tend to fall—not as much as Tom's for Jack, but more than you may think. A vast web of these mutual esteem relationships surrounds the individual. And a GEL or GET point can trigger a chain of mode drops all through this web.*

The general esteem curve may be represented by the following table.

General Esteem	*Mode of Function*
Pity; contempt, ridicule	Identic to Reactive
Blame; regret; desire to punish	Reactive
Distrust; suspicion	Reactive to Uniordinal
Disagreement; misunderstanding	Uniordinal
Mutual understanding; trust	Multiordinal
Synergic alignment of goals; empathic communion; totaltalk	Synergic
Excessive trust; dependency; desire to propitiate	Uniordinal to Reactive
Awe; abasement; complete submission, absolute obedience	Identic

There is one important basic difference between general esteem and self-esteem. General esteem involves a two-way relationship. It implies not only the esteem of the individual for some entity, but also the esteem of that entity for the individual. This has two important practical applications.

---

*In the early phase of the development of synergetics, this general esteem web was recognized, but its vital importance was not appreciated. Our perspective was focused on Individual Synergetics and we tended to overestimate the capacity of the individual to free himself completely from the reactive entanglements of the general esteem web. This problem will be dealt with later.

a. Whenever either the individual or the entity experiences a GET or GEL point, he may react. His reaction evokes a GET or GEL point in the other, who reacts in turn. A chain reaction tends to occur—A reacts, then B reacts, then A reacts again, and so on. Such chain reactions are called twangles in synergetics. They can go on indefinitely.

b. Because of the general esteem web, the individual is influenced by the reactions of all the entities with which he has a relationship. This means that his total dysergy load is much greater than that produced by his personal impedances. Even when he avoids or prevents a twangle, he may often be subject to irrational demands and restrictions that are difficult to manage, even in the synergic mode.

Still, by rational re-evaluation of GEL and GET point reactions, it is possible for the individual to clear his own reactions and to neutralize those of others. This makes it that much easier to achieve the synergic mode.

3. A pleasure thwart. The desire for pleasant experiences is a natural and basic human desire. Pleasant experiences in themselves do no harm; on the contrary, they usually promote the survival of the individual or the human race. A built-in desire for pleasure is sound biological design.

Unfortunately, the desire for pleasure is often thwarted in one way or another. When this happens, the individual may react. The reaction usually takes the form of an intensified effort to achieve pleasure.

The most intense form of pleasure is sexual, of course. Hence sexual pleasure thwarts are a very powerful source of Identic and Reactive patterns. The influence of Freud, Reich, and others has helped considerably in the reduction of such patterns, although Utopia has hardly yet been achieved. Unfortunately, such practices as the commercial exploitation of women as sex objects has been a strongly negative influence in this respect.

There are two particular reactions involving pleasure that are of interest. Those are the *pleasure circuit* and the *pain drain*.

The pleasure circuit starts with a pleasure thwart. The individual reacts. The reaction takes the form of an intensified desire for pleasure. This desire seeks an outlet. Being reactive, it tends to be excessive or extreme. This may do harm, or threaten to do harm. The harm, or threat of harm, in turn produces a reaction. The reaction operates to restrict the pursuit of pleasure. This restriction operates as a pleasure thwart. This second pleasure thwart reinforces the first. This produces a further intensification of the desire for pleasure. And so on and on in a circuit.

The pleasure circuit is at the root of many a bad habit such as excessive smoking or excessive drinking. It is also found in reactive morality such as the "hellfire punishment for yielding to temptation" of some religions. It is a beaut of a dysergy trap.

The *pain drain* originates not in a pleasure thwart but in a pain threat (see below). When a pain threat occurs, the individual reacts. This reaction may take the form of avoidance. The opposite of pain is pleasure. A reactive outlet for pain avoidance, therefore, is an intensified desire for pleasure. Pleasure thus serves as a "drain" for avoiding pain. Pain drains may be at the root of alcoholic binges, excessive reading of escape literature, excessive watching of TV, etc. An excessive, reactive interest in politics or religion may be other examples. Still another may be a vicarious delight in witnessing the pain of others, or inflicting pain as an anonymous member of a crowd as in mob violence or lynching. The hate ideologies such as Nazism, Stalinism, anti-Communism, anti-Semitism, racism, etc. achieve strength because they provide safe pain drains.

4. A pain threat. Pain is the alarm signal of the body. It tells the mind that the body is about to be damaged. It produces instantly a number of reflexes whose net effect is to move the body away from the source of injury. It also brings into action a number of bodily mechanisms to repair the damage.

We are here primarily concerned not with pain itself, but with *pain threats.* When physical pain itself occurs, it may be because of injury or disease; if severe or prolonged or accompanied by other signs such as fever, a physician should be consulted. This is outside the scope of synergetics.

Pain threats are another matter. The organism "remembers," often unconsciously, painful incidents of the past and whenever a stimulus similar to some part of such an incident occurs, the organism may perceive this as a pain threat.

The response to a pain threat is a *reaction*—something that "worked" in the past (i.e., that was followed by reduction or elimination of pain). These reactions may be of two types: mental reactions, and bodily reactions. (Often both are evoked.)

Some examples of mental reactions are:

Attenuation—A "turning down" or "turning off" of consciousness. The individual simply does not notice consciously some of the processes that are going on. If severe, the individual may actually faint.

Emotional states—Any of a variety of emotions experienced during a painful incident of the past may be turned on again, often with no apparent relation to the present situation—fear, anger, grief, apathy, appeal for sympathy, etc. The "charge" or intensity of the restimulated emotion may vary from light to strong. Sometimes the emotion has no "name" and is difficult to describe in words.

Confusion—Ideas, images, recollections, perceptions, etc. may occur in a random fashion, with the individual experiencing difficulty in sorting them out and establishing relationships among them in a rational way.

Memory blocks—Frequently, a blockage against recall of some aspect of the previous incident may occur.

The bodily reactions may include unconscious or semiconscious muscle tensings, semiconscious movements, or various sensations in different parts of the body. There may also be involuntary physiological reactions such as sweating, increased blood pressure or heart rate, increased breathing, etc. Often these bodily reactions form a coordinated pattern with a purpose; for example:

Avoidance reaction—This is a set of muscle tensings and movements to withdraw from the pain threat.

Protection reaction—This takes the form of an increased sensitivity or tendency to act—a general tensing of various muscles, with nervousness or trembling, increased sweating, etc.

Attack reaction—This may take the form of increased body activity, flushing of the face, increased heart rate, etc. The body is mobilized to attack and destroy the pain threat.

Submit reaction—This may seem like an odd way to deal with a pain threat, but its purpose is to "propitiate" the source of pain in the hope that the source will reduce or eliminate the pain. The reaction may include a general relaxation of muscles, a readiness to conform to external forces, and a tired feeling.

These are only a few of the different types of mental and bodily reactions that may occur to a pain threat. As noted previously, mental and bodily reactions usually occur together. Also as noted, reaction to a pain threat may involve an excessive pursuit of pleasure (the pain drain).

When a motive is tracked down to a pain threat reaction, there are two ways to deal with it. The first is simply to confront the pain threat objectively and rationally, and to devise a synergic response to replace the reaction. The second is to use a technique called Incident Recall. This will be discussed later.

5. A survival threat or loss. Any event, situation, person, group, experience, etc. that is perceived as a threat to the survival of the individual or his symbiotes may evoke reactions that are actually in large part recordings of similar incidents in the past. Here, it is desirable to distinguish two cases: one in which the survival threat is immediate, requiring emergency action to handle it, and one in which the danger, though real, is not immediate or strong enough to require emergency action. Clearly, the first case requires a concentration of mind and effort directly on the threat itself. However, this will not ordinarily be encountered in a synergetic session.

When a motive is tracked down to a survival threat reaction, again there are two options: replace it with a synergic response, and use Incident Recall.

In the life history of a person, there may also occur incidents in which survival is actually "lost" in some way. A parent or other close relative may die, or be severely injured, or lose a job or otherwise suffer economically. The person himself may lose a limb, or suffer other bodily injury, or be compelled to give up something he needs or deeply wants.

A survival loss of any kind may produce a strong attenuation in the mind of the individual, in some particular domain of function. Very often a *self-invalidation* occurs—a "turn-off" of the creative evolution of the individual—he stops developing in some area. He may experience deep grief, terror, anguish, or despair. He may blame some person or group and form a deep and abiding hatred. He may even blame himself and feel a deep and abiding remorse or guilt.

When a motive is tracked down to a survival loss reaction, the tracker again has two options. The first, as before, is to replace it with a synergic response. The second is to use a technique called Creative Tracking. This is described in chapter 10.

6. A value threat. When we are young, our minds lack data. We are "fed" data by simple interactions with the environment, by our parents, by siblings and playmates, by relatives, by teachers at school, etc. In the absence of reason to the contrary, we accept this data as "valid." Much of it is. But much isn't.

Some of this data has to do with values. The young mind has the inherent capacity to assign worth to various aspects of experience, but it depends, to a large extent, on data from outside for the basic values according to which worth is assigned. If mom or dad says (or implies by action) that something is "good," the child tends to accept this, and thereafter to use this datum as a basis for evaluations. Similarly, if mom or dad says, or implies by action, that something is "bad," the child accepts this.

This vast early inflow of data and values shapes our minds profoundly (a fact not unknown to the twig-benders). The mind has marvelous cross-checking devices, but it cannot be proof against false data, and it is especially vulnerable to dysergic values. When such data are "stamped" at the input as "true" or "good" or "right" or in some sense "valid," and the mind knows nothing to the contrary, the mind accepts the data and thereafter uses it to validate other data and to assign worth-values. The twig has been bent.

When a motive is tracked to a value threat reaction, the tracker can replace the reaction with a synergic response. This takes care of the immediate problem. But if the value is itself dysergic, sooner or later it will produce more dysergy.

Thus, it is often worthwhile, when a value threat reaction is encountered and replaced, to re-evaluate the value itself. This leads to another mental program—the Value-Interest-Perspective (VIP) Triangle. This is discussed later.

Motives, then, can almost always be tracked down to one of the following:

SEL or SET Point Reactions
GEL or GET Point Reactions
Pleasure Thwart Reactions
Pain Threat Reactions
Survival Threat or Loss Reactions
Value Threat Reactions

It is convenient to have one term to describe these basic motivations. We refer to them as *determinants*. In a sense, they are the "push buttons" of the mind.

The *attitude* side of the BAM Triangle refers to the *mental or bodily set* of the individual. An attitude consists of two parts: *expectations* and *anticipations.*

The beliefs a person accepts and the motives he adopts determine his attitudes. Based on these beliefs and motives, he *expects* certain things to happen. If a woman believes a man likes her, she expects that he may make a pass at her. If a child believes he has done something that his parents regard as wrong, he expects that he may be punished.

The expectation is a prediction of what may happen. But the mind does more than just predict. It also *anticipates* what it expects. The child cringes before the expected blow. The woman avoids situations or actions that would facilitate the pass (or the opposite, if she is so inclined).

Expectations and anticipations—the components of an attitude—are nonverbal and often difficult to describe in words. They also are, to a large extent at least, determined by beliefs and motives. For this reason, less attention is sometimes paid to the attitude side of the BAM Triangle than to the other two sides. However, it should not be ignored; examination and evaluation of attitudes often lead to useful insights and provide the tracker with an awareness of processes he usually does not know about.

Other Programs

Tracking, using the CEDA Sequence and the BAM Triangle, is a very powerful technique for turning on the synergic mode in a synergetic session. There are a number of other programs and devices that may also be useful, and indeed the tracker soon discovers or invents programs of his own, which he adds to his evolving program library. Here are a few that I have found useful.

1. Directives. A directive is a program for clearly defining a track and is often used to start a session. It consists of the following parts:

a. A *goal.*
b. An estimate of the *value* or values of achieving that goal.
c. A *program* for achieving the goal.

Thus, for example, the tracker might formulate the following directive:

Goal   To achieve a synergic grasp of tracking.
Value   (1) To clarify thinking.
          (2) To turn on the synergic mode.
Program (1) Re-read this chapter.
          (2) Try a synergetic session, using the CEDA sequence and
              the BAM Triangle.
          (3) Evaluate the experience.

The use of directives introduces a high degree of precision into
tracking. The tracker forms a clear idea of where he is going, why he
wants to get there, and how he plans to do so.

Directives are also of value when, as sometimes happens, the
tracker gets "off track." By simply reviewing the directive, the tracker
quickly gets back on track.

2. The ART Sequence. This consists of three operations: Accept,
Reject, Transcend. These are applied in turn to any belief, viewpoint,
situation, individual, group, etc.

Let us suppose, for example, that the tracker has a belief that he is
clumsy with his hands. First, he *accepts* the belief, and explores its
implications. Then, he *rejects* the belief—"It ain't necessarily so"—
and examines the implications of so doing. Usually, one or more new
viewpoints will emerge, for example, a viewpoint that motor skills are
actions that can be learned to varying degrees of effectiveness, that
most people are usually clumsy at first. Then he *transcends* the belief.
He may do this in either of two ways:

   a. By searching for a viewpoint in which he is both clumsy and
      nonclumsy. This may seem absurd at first, but try it and see
      what happens. H'mmm...clumsiness is relative, and depends
      on the *standard* that is being applied. Picking up a marble with
      toes is clumsy compared to fingers, but rather skillful con-
      sidering the structure of toes.
   b. By searching for a viewpoint in which he is neither clumsy nor
      nonclumsy. From the standpoint of getting a job done, it really
      doesn't matter whether one is clumsy or not as long as the job
      gets done.

The ART Sequence is useful in many ways and for many reasons.
One of its chief values is that it quickly generates a Multiordinal
awareness. It is especially good at bringing to attention new and
previously unconsidered aspects of a situation. It is a powerful tool for

clearing dysergic BAMs. Used repeatedly, it enables the tracker to grasp complex situations from a variety of viewpoints and to estimate rapidly the effects of various lines of action upon each of these viewpoints. This can be very useful in influencing a group of people of diverse interests toward synergic action. I have watched syngeneers do this on occasion and have been amazed. Suddenly a difficult problem gets solved and nobody really knows how it happened, but everyone is pleased. Only another syngeneer who is himself multiordinally aware can appreciate what has been done.

3. The Value-Interest-Perspective Triangle (VIP Triangle). This is a program for dealing with dysergic values and false and erroneous data that we have accumulated since infancy.

By *value* is meant the worth assigned to things, ideas, individuals, groups, activities, etc.

By *interest* is meant a readiness to pay attention to or be concerned about objects of value.

By *perspective* is meant a mental frame of reference in terms of which contents are identified, relationships established, and values assigned.

An example of a VIP might be the following, toward "freedom."

Value: It enables me to do what I want or choose.
Interest: Orients me favorably toward things that promote my freedom and unfavorably toward things that impede it.
Perspective: Freedom involves the absence or minimization of external demands and constraints.

As with all tracking, once a VIP has been stated, the tracker then asks himself why he accepts it, what other VIPs are possible, etc. Thus: "Why do I want to do as I want or choose? Could this be, in part at least, a reaction to an incident or a chronic situation in which I was prevented from doing what I wanted?" Etc.

Few people realize the extent to which their consciousness is limited and its contents are determined by their VIPs. Few people have ever undertaken an honest and searching re-examination and re-evaluation of their VIPs. We cling to them with a tenacity that suggests that in all too many cases they are reactive in origin. The oak has forgotten how the twig was bent.

Perhaps one reason we cling so strongly to our VIPs is that intuitively we feel it is dangerous to question them. In this, we may be right, for VIPs are often shared with others in groups to which we belong. To question openly the VIPs of our groups or society might well bring down upon us a collective wrath that an individual could not withstand.

But a synergetic session is a *private* affair. And tracking in no way compels anyone to abandon a VIP if, on honest and searching re-examination and re-evaluation, he finds it still one that he wants to keep. On the other hand, if you were bent as a twig, wouldn't it be more comfortable to straighten up?

And unless you have done an honest, searching, and systematic re-examination of your VIPs, how can you be sure they are clear of dysergy? Or that many of your troubles and personal impedances are not at least in part traceable to dysergic VIPs?

4. Incident Recall. In examining a belief, it is often helpful to recall a specific incident in which that belief was first formed. The same applies to a reactive motive.

In Incident Recall, the tracker tries to recall, in chronological order, the significant details of a particular incident, exactly as it happened. This is done as objectively as possible. At first, the tracker may recall only isolated fragments of the incident; but if he goes over it repeatedly, each time focussing on a different aspect of the incident—what he saw, what he thought at the time, what his emotional response was, etc.—more and more details will come into view.

Incident Recall is of value for two main reasons: first, the effort to recall and to view objectively a past incident tends to reduce the emotional charge associated with that incident; and second, it enables the tracker to review and re-evaluate the data and interpretations on which a belief is based, or to clear a reactive motive. The repetition of sweeps through the incident seems to be largely responsible for the effects produced.

Conclusion—and Beginning

These, then, are some of the tools of tracking, the basic technique of Individual Synergetics. There are many other programs, of course, and you will soon find that you can formulate and apply your own. In the following a number of these programs will be described. But, as of now, the basic ideas and techniques to enable a person so inclined to *start tracking* on his own have been presented. All that is needed is a decision to do so.

To any who so decide, I'd like to present two points for your CEDA:

1. A number of ideas and tools have been described in the foregoing. Anyone who has read thus far has stored them in his "memory banks." But there is a world of difference between conceptual grasp of an idea and a synergic grasp of the same idea. The *concrete experience* of a few tracking sessions will provide you with a new basis for interpreting and evaluating what has been presented thus far, and for integrating these ideas and tools with your own knowledge and experience. Therefore, I respectfully suggest that, after a few sessions, you re-read this book up to this point. I believe you will find it well worthwhile.

2. Tracking is *not* a mechanical ritual, and no matter how many acronyms (like CEDA and BAM and ART and VIP) we have used, these are only mnemonic aids to facilitate succinct expression and recall. There is a *spirit* of tracking that can only come from the tracker himself and that transcends by far all the ideas and tools that have been labelled "tracking." It is a spirit of honest, open-minded, probing inquiry—the search for the truth about oneself. This spirit is vital to effective tracking, and it is precious.

We begin to realize that our brains are the most complex and self-determining things in the known universe...If this property of complexity could somehow be transformed into visible brightness so that it would stand forth more clearly to our senses, the biological world would become a walking field of light compared to the physical world...An earthworm would be a beacon, a dog would be a city of light, and human beings would stand out like blazing suns of complexity, flashing bursts of meaning to each other through the dull night of the physical world between. We would hurt each other's eyes. Look at the haloed heads of your rare and complex companions. Is it not so?

————John Platt (1)

CHAPTER 8

THE SYNERGIC MODE

As far as is now known, the synergic mode of function is available to every human being no matter what he has done or what has been done to him. It is one of the sublime prerogatives of being human.

One might well wonder: if the synergic mode is so available and so rewarding, why is it so rare? Why don't more people use it, especially those in positions of leadership and influence? Why doesn't everyone use it as much as possible?

No doubt there are many reasons for this, but two in particular seem important. The first is ignorance—our lack of knowledge about how the mind works. This is not to say that nothing is known—on the contrary. But compared to what remains to be discovered, our position is very much like that described by Isaac Newton, who opined that he had found a few pretty pebbles on the beach but that the vast Ocean of Truth lay unexplored before him. It would be wonderful if we knew enough and had the facilities enough to create a vast communication network uniting the planet with stations to which every human being could come and be turned on, so that one vast, world-wide surge of synergy would sweep joyfully into the minds of everyone, and we would suddenly know, with glad amazement, that humanity is one.

But we do not now have that knowledge; and even if we did, we do not have the facilities. Still, it is a long-range goal toward which we can move, a step at a time.

We do not know enough. But we do know that tracking, in synergetic sessions, turns on the synergic mode. It doesn't always work, and it may not work for everyone; but it happens often enough to give us hope that we are on the right track. Furthermore, when it doesn't work, analysis shows that either tracking has not been done properly or the dysergy load of the individual is too great.

The second reason for the rareness of the synergic mode is dysergy. There is some dysergy within the mind of almost everyone. Indeed, this is acknowledged by nearly everyone, though different words are used to describe it. We all have mental patterns* that, like static on a radio or "snow" on TV or bugs in a computer program, produce errors in perception, confusion in thinking, memory blocks, blue moods, bad habits, etc.

But quite apart from individual dysergy, there is a vast amount of dysergy in society. There are, first of all, the impedances of others with whom we are closely associated that at best make things more difficult, and at worst lead to those chain reactions we call "twangles." Beyond this there are dysergy patterns in social processes that lead to war, crime, poverty, racism, sexism, etc. These we call *sociodynes;* and their impact on the individual is subtle and profound.

Thus, everyone is weighed down with a heavy dysergy load consisting of personal impedances, the impedances of others, twangles, sociodynes, and other forms of dysergy. The sensible thing to do would be to reduce this dysergy load as much as possible; but most of us are too busy, or too involved with the *consequences* of dysergy—the worries, the frustrations, the problems of everyday life—to take direct action. Still, a reduction of one's dysergy load would be desirable in its own right.

From a synergic standpoint, there is another reason for reducing dysergy: *it deprives us of the delights of the synergic mode.* To one who aspires to synergy, what was once tolerable becomes an obstacle that must be removed.

In ensuing chapters, we will describe some techniques for reducing

*We refer to such patterns as "impedances" in synergetics because they impede mental function.

and eliminating dysergy. In this chapter, we present additional techniques for generating synergy. These can all be used in synergetic sessions and to a certain extent outside as well.

Tuning

Chance favors the prepared mind.

——Louis Pasteur

When a person switches on a radio, he usually has to turn a knob in order to dial in the station he wants. As he gets closer and closer to the station, the sound grows louder and louder until it reaches a peak. If he turns too far, the sound fades again. By careful adjustment, the maximum level of sound is reached. The station has been tuned in.

So it is with the world about us, in any situation. In the welter of data available to our sense organs and other receptors, some are synergic and some are not. A datum is synergic if it challenges the mind, facilitates a goal, promotes a value, excites wonder or joyful anticipation, etc. A beautiful sunrise, good music, a courageous act, an interesting book, an exciting companion...it is difficult to give an adequate definition in words to a synergic datum or data set, but the mind knows synergy when it feels it.

Some data are not synergic, but not dysergic either. They are more or less neutral as far as synergy is concerned. But just as a desired radio station is found on a small part of the dial, so are synergic data vastly outnumbered by neutral data.

But we can tune in synergic data by a mental act. By searching for synergic data and homing in on it wherever it is to be found, we can increase the synergic input to our minds.

This is very simple and elementary, and the wonder is that it is not done more often. It is reasonable to assume, and easy to demonstrate, that a mind is more synergic when it receives synergic data. This is especially important when we realize that a large part of the contents occupying the mind at any moment come from sensory inputs of the moment. Yet, in my experience at least, most people do not tune very often. Perhaps the reason is that it has never occurred to them to do so.

Yet, simple though it is, tuning is not always easy. In a welter of neutral and dysergic data clamoring for attention, a synergic datum is often not immediately apparent. Indeed, to a certain extent, synergy, like beauty, is in the eye of the beholder. Tuning to synergy requires a prepared mind, one sensitive to synergy and ready to home in on it when it is detected to the exclusion of other matters.

Tuning is more difficult when dysergy is present. Yet, even here—perhaps especially here—tuning is very useful. In the previous chapter we spoke of the general esteem web. If your esteem for another falls, or vice versa, you may feel an urge to react. Your reaction may trigger a chain reaction—a twangle. But often it is possible to resist that urge and to tune instead. This does not mean denying the reactive elements in another's behavior. But each person is also a potential synergy source. Look for the synergy in him, and tune to that instead. When this is done, it often happens that a mode rise is evoked.

Tuning can be done in various ways as long as its essential feature of detecting and homing in on synergy is preserved. One simple way to do it is by a simple program: (1) Search for synergy, (2) Amplify the synergy, (3) Reduce the noise.

Finally, it is worth pointing out an intriguing feature of synergetic tools: *they tend to be synergic to each other.* A person who is tracking finds it easier to tune than when he is not. Conversely, tuning facilitates tracking. The same can be said of Panview, described below.

Panview

The average person characteristically views incoming data from one perspective only—his own. Since he views data from the relatively restricted mind band of ordinary consciousness, even here his perspective is smaller than that of a synergic being. Now the individual's own perspective is important; it should not be ignored or repressed.* But it is always possible to expand awareness to include the perspectives of others.

Panview is such an act. It is the mental operation of deliberately viewing a set of contents from two or more different viewpoints. Indeed, both viewpoints can be the individual's own. Each of us sees things differently at different times. There is nothing to prevent us

---

*Here we would like to propose the outlook of *synergic altruism*. This does *not* mean self-sacrifice or "selflessness," which the term "altruism" sometimes connotes. It means actively helping others without sacrificing oneself. Synergy is that electric surge of the mind that occurs when it zooms from "either-or" to "both-and."

from seeing things differently at the same time. In one aspect, a problem is an obstacle to be overcome; in another, it is an opportunity to learn.

Panview expands awareness and sets the mind for synergy. In addition, it facilitates tracking, especially the Consider phase of the CEDA Sequence. Sometimes it is not easy to formulate two possible lines of action. Panview makes this easier.

There are other uses:

1. Panview enables the individual to gain a better understanding of what he is looking at. From any single perspective there are always features of an object that are hidden or obscure. The view of a forest from the air, the state of the floor beneath a carpet, the inside of another person's mind—these are aspects of familiar objects we seldom see or know. Of course, it is not feasible to see everything, to know all the angles. But it is always possible to see more, from a different perspective, and the more we see, the better we know, and the freer we are from distortions and illusions inherent in any one perspective.

2. Panview is helpful in human relations. The one who panviews a situation sees it not only as it appears to him, but also as it appears to others. He is therefore in a position to understand how two people can differently interpret the same event. This enables him to take positive action relative to the event, which will promote the interests of all concerned. And he is better able to learn from the experiences of others.

3. Panview promotes the registration of data in memory, and thereby renders them more accessible to recall.

4. Applied to words, panview produces semantic awareness—the realization that words are not identical to the things they represent, that the same words may have different meanings to different people, that the meaning of a word changes with context, etc.

We live so largely in a verbal world
We lie so limply in a verbal whirl
Words with branched meanings, within contextual clusters
Words in chance movement, while we gaze entranced
At a sky shaken by a giant hand
Stars in a cosmic dance...
Gyrating
Rhythmically

Endlessly grand
Settling
Majestically
Into the land...
Until significance musters
And we understand
As the banner of Truth
Is
(Silently)
Unfurled.

It is amazing how much semantic awareness can improve a person's ability to orient effectively to the "real world" rather than to the cage of words that circumscribes the vision of so many, so much.

Thrillbeam

During the flow of input data, or of ideas in conscious thought, there occur from time to time *opportunities to promote synergy by action.* The ripe moment to persuade husband or wife to do something both know needs doing; the creative compromise to spring on contending parties when both are weary of the struggle; the key idea that simplifies and clarifies a confused situation; the shift of perspective that enables you to combine two seemingly conflicting viewpoints—these are examples of *synergy points* where much may be gained with little effort.

But synergy points do not just happen. They occur to minds that are set to detect them and ready to act instantly before the ripe moment passes. They occur especially to minds in which investments of time, energy, and thought are regularly made in things that tend to generate synergy—in other words, in *synergic potentials.*

Such investments are more readily made when emotion is mobilized in support of thought and action. Thrillbeam is a tool that accomplishes this.

In Thrillbeam, the individual orients to synergic potentials available to him. As he does so, he also specifies the tremendous potential value to him of maximum use of these potentials.

Thus, for example, the synergic application of a synergetic tool (like tracking or panview) to situations in which they are applicable is a synergic potential. Note the emphasis on synergy of application—the skill and loving care with which the tool is used and the fitness of the tool to the situation are as important as the tool itself. And for someone who longs to experiences the synergic mode, to achieve at the best of which he is capable, this synergic potential is indeed of tremendous potential value.

Again, each individual possesses inside his head a supercomputer of fabulous powers. This is not a matter for pride—it was *given* to him, whether he deserved it or not. Few use it at more than a fraction of its capacity, and there are latent abilities within it that have never been activated. This magnificent instrument is the culmination of billions of years of evolutionary development, an evolution that is still going on. Here is a synergic potential of enormous value.

And *knowledge* of this supercomputer is another synergic potential.

In a similar way, other human beings are synergic potentials toward whom the syngeneer can direct a Thrillbeam. Indeed, with a little thought, the syngeneer can list many other synergic potentials on which to Thrillbeam. As he does so, he matter-of-factly specifies the potential value of utilizing the synergic potential. Sooner or later, an emotional mobilization occurs—usually one of joyful, enthusiastic anticipation. Look-feel synergy has emerged.

A mind set by Thrillbeam is observably more apt to detect and act upon synergy points than one that is not. Sometimes synergy points and synergic actions occur so rapidly that the individual just blurs. The effect is rather pleasant.

Clearlook

Children, immature adults and animals "identify."
Whenever a person reacts to a new or changing situation
as if it were an old and unchanging one, he or she is said
to be identifying.

——A.E. Van Vogt (2)

Properly used, Thrillbeam is a powerful technique. But it is best used in moderation, lest the syngeneer find he is riding a manic. The result, later, may be the blahs. Fortunately, an antidote exists for the latter. It is called Clearlook.

The average person receives input data passively and imprecisely.

Little effort is given to accurate, critical appraisal of the data in terms of probable truth value and probable survival value. This passive perception is extensively exploited by modern advertising and propaganda. Love that soap!

Of equal importance is the distortion of input data by wishful or fearful thinking. Through this distortion a large amount of false or inaccurate data enters the mind. These data lead to confused thinking, erroneous predictions, emotional conflicts, inept actions, and much more.

Another major source of error is *semantic identification.* By this is meant an unconscious tendency to *identify* a word or verbal expression with the *referent* it represents. A typical example is the tendency to identify the *name* of a person with the person himself. Name-calling such as calling a person a "bastard" or a "Communist" is another example. The response of many individuals to name-calling indicates a strong emotional reaction to a semantic identification. Being called a bastard or a Communist does not make it so. But the response of many individuals indicates fear that it will be so if they do not attack, verbally or otherwise, the person calling the name. Further, if they do attack it will "prove" the name "false."

Semantic identification is highly exploited by advertisers and propagandists. People are led to identify the brand name with the product. They end up buying labels. The ideas and techniques of general semantics* are highly recommended here. Training in general semantic techniques enables the individual to reduce or eliminate such semantic identifications with considerable benefit.

Clearlook is a powerful tool that not only clears such errors, but also provides an increased capacity for handling input data. It consists of three operations, as follows:

1. Objective, precise appraisal (OPA). The individual carefully separates what he wants to see or fears to see from what is actually there. In other words, his look is objective.

In addition, the individual appraises the data as precisely as possible. He measures or carefully estimates "how much" or "how many." He specifies and distinguishes the different properties of input data. The degree of precision may vary and should be as sharp as is needed or desirable for his goals—in other words, he is also precise in the degree of precision used.

---

*See *Science and Sanity* by A.L. Korzybski. (3)

2. Semantic discrimination. The individual continuously makes a distinction between word and referent, between verbal map and territory represented. He is continuously aware that the verbal map is not 100% accurate and is almost certainly incomplete. He "mentally punctuates" verbal data from time to time with "etc." to remind himself of *things left out* of the verbal description. He is continuously aware that a word may mean different things to different people and even to the same person at different times.

Semantic discrimination gives the individual a considerable advantage, since it orients him to the reality rather than to the verbal map and enables him to detect and to eliminate errors he would otherwise miss. It promotes an extraordinarily precise and synergic interaction with *events,* independently of their verbal representation. This synergic interaction often evokes the phenomenon of "flash-grasp" of complex situations.

Semantic discrimination also immunizes the individual against advertising and propaganda.

3. Information Source. Every individual is confronted in the course of his life with an enormous variety of input data. In order to handle the data efficiently, he assigns it to *categories.* This organizes the data and enables him to respond economically and in an organized way.

Unfortunately, once a datum has been assigned to a category, there is a tendency to ignore other data associated with the first datum and to act entirely in terms of the category. This is especially true of bureaucracies (public and private): once a person has been classified, he is treated in terms of his pigeonhole, without regard for his needs and potentials as a real person. It is also true of new ideas; the characteristic response of rigid minds is to place the idea into an already known category so that it can be safely ignored. This is as true today as it was in the time of Galileo, and it can be confidently predicted that synergetics will meet this fate in some circles.

But there are almost always novel elements—*emergents*—among the input data stream. To regard an input data set, a person, a situation, etc., as an Information Source is to regard the situation as an opportunity to learn something new. It opens the mind. And clearly a person with an Information Source orientation is more likely to learn something new than one who does not.

Clearlook, then, consists of these three operations:

Objective, precise appraisal (OPA)
Semantic discrimination
Information Source

They can be done in sequence, but also—with practice—all at once.

Thrillbeam and Clearlook form a synergic pair. Thrillbeam mobilizes emotion in the service of thought and apt action; Clearlook stabilizes the look-feel synergy that thus emerges. Together they make quite a team.

In a similar way, Tuning and Panview form a synergic pair. And when these two synergic pairs are synergically combined . . . .

Synergic Goals and Aims

> *Evenness of beat would . . . lead to rigidity of music. It is the fluctuation in the meter, the agogic molding which "disturb" the frozen pattern of synchronization and breathe life into the work of art. It is this which creates the inner motion, makes contact with the listener and produces the . . . audible event.*

> ——Fritz Winckel (4)

The basis for synergy in time is purposive activity. A clearly defined goal, adopted freely by the individual to promote some value, and pursued despite obstacles and distractions generates a flow of synergy in the mind. A variety of mental processes are activated, coordinated, and focused about the goal. Problems necessary to its achievement are posed, data necessary to solve these problems are acquired, stored knowledge and skills spring to mind. Where indicated, the goal is broken up into a sequence of subgoals, each leading a step closer to the desired end. The mind does all these things smoothly, automatically. Its natural synergy is made manifest in the process.

The choice and pursuit of a goal is relatively easy in a synergetic session, or in any situation in which the individual is dominant. In everyday life, however, such situations are not common. At work, in school, or in the army, we are confronted by situations in which goals are set by others. We follow orders. Even the goal-setters are not free, but have others above them in command hierarchies; and the one on top is a prisoner of his roles, of demands, expectations, and constraints that clamor for his attention from all sides.

The *synergic goal* provides a neat way around this. By a synergic goal is meant one that promotes two or more values at once. "My way home goes by your house." It is, in other words, a multipurposed goal.

The song of a radio singer is carried first by sound waves to a microphone, then by electric currents to a radio transmitter, then by electromagnetic waves through space to a radio receiver, then by electric currents again to a speaker, then by sound waves again to a listener. The song is a signal superimposed on a number of different physical carriers.

So it is with a synergic goal. When a person receives an order, he accepts an external goal that is not his own. But in carrying it out, he can almost always superimpose on the actions he performs *another* goal, a goal of his own. The external goal is a carrier for his personal goal. And, as long as the two do not conflict, the total goal is a synergic one.*

It frequently helps to design synergic goals in a synergetic session. Thus a synergetic session becomes, so to speak, a period in which the individual achieves synergic command of his life.

A goal is specific and, once it has been achieved, activity ends as far as the goal is concerned. Goals sometimes form hierarchies, with subgoals having to be achieved first and supergoals above, as illustrated on the next page.

However, there are some purposes that are never finally achieved. Learning about a particular subject is one; loving a person is another. Such purposes might be called *aims,* to distinguish them from goals. An aim can be regarded as an infinite source of goals. And a *synergic aim*—one focused so as to avoid impeding the rational or synergic goals and aims of other parties—provides a powerful basis for a temporal organization of one's being.

A person who, in synergetic sessions, has adopted a set of synergic aims achieves thereby a state of synergic becoming. He is never at a loss for things to do. His whole life is focused on synergy. We call such a set of synergic aims a Synergic Basis.

A person who effectively implements a Synergic Basis experiences the synergic mode more and more often. The time eventually comes

---

*This is not to say that we advocate or accept a system based on hierarchy and authority such as now prevails in every nation on earth. We live in a *very* dysergic world. The "enlightened" twentieth century is the bloodiest of all time; and a nuclear sword of Damocles still hangs by a thread over humankind. But the struggle to change the system will be long and difficult, and meanwhile people have gotten into a bad habit: eating.

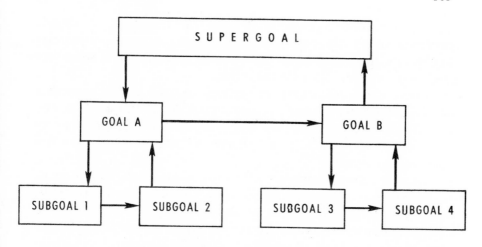

GOAL HIERARCHIES

when he is synergic almost continuously. He may experience mode drops occasionally, but he quickly traverses to synergy again. Such a person is said to have *stabilized in the synergic mode.* *

A Synergic Basis, it should be pointed out, is not alone sufficient for stabilizing in the synergic mode. To achieve this, it is necessary to eliminate impedances. But more on this later.

Drivemaster

Each synergetic tool tends to generate synergy whenever it is properly used. This raises an interesting question: what happens when synergetic tools are synergically combined?

The effect cannot be described adequately in words; it must first be experienced to be understood. We have already encountered several of these "synergic combos"—CEDA and BAM, Tuning and Panview, Thrillbeam and Clearlook. Furthermore, with CEDA as base, synergic combos are readily formed with all synergetic tools.

Here is one more: Drivemaster. It consists of three tools:

1. Goal CEDA
2. Directive
3. Totalact

---

*Such individuals are sometimes called *stables*. However, they dislike the term and prefer not to call undue attention to themselves; they are in no sense "supermen" or "superwomen."

Totalact involves focusing on the goal, and exerting an all-out effort to achieve it. The aiming point for the effort-focus is "Flexibility as to means, fixity as to goal." This means synergic use of whatever means are available with holistic effort.

Synergic Use. If, as sometimes happens, use of a particular means unavoidably harms another, compensation for harm done is accomplished as soon as possible. In the synergic mode, however, the individual usually is able to anticipate and prevent such harm by quick adjustments.

Totalact involves precision of effort, not waste effort. The effort is *placed* so as to achieve optimum gain, *timed* so as to achieve optimum effect. Where efforts are combined, the *order* in which the efforts are made is chosen for optimum economy of effort. To put it succinctly, Totalact involves TOP effort (*T*ime, *O*rder, *P*lacement).*

Drivemaster is not a tool that should be used continuously, however, In most situations, a goal CEDA followed by a Directive is sufficient. Drivemaster is best reserved for special situations in which its use is warranted.

Conclusion

In this chapter, a number of tools have been described which, when properly used, generate synergy. For convenience, they are again outlined below:

1. *Tuning*
   a. Search for synergy
   b. Amplify the synergy
   c. Reduce the noise
2. *Panview*—view contents from two or more perspectives.
3. *Thrillbeam*
   a. Orient to synergic potentials
   b. Specify value of using those potentials
4. *Clearlook*
   a. Objective, precise appraisal (OPA)
   b. Semantic discrimination
   c. Information Source

---

*Long before now, some readers may have concluded that I am an acronym addict! The acronyms are not intended to mystify, but to act as aids to memory—mnemonic devices—as well as for succinctness of communication.

5. *Synergic goals*—use externally imposed goals as carriers for personal goals.
6. *Synergic aims*
   a. An aim is a (virtually infinite) *source* of goals.
   b. A *synergic* aim is focused both to achieve something and to avoid impeding others.
   c. A *set* of synergic aims—a Synergic Basis—is necessary but not sufficient for stabilizing in the synergic mode.
7. *Drivemaster*
   a. Goal CEDA
   b. Directive
   c. Totalact
      (1) Flexibility as to means, fixity as to goal
      (2) TOP effort (Timing, Order, Placement)

These tools can be learned and used relatively easily in a synergetic session. In life situations, however, it is usually more difficult.

There are a number of reasons for this, which will be considered in later chapters. Basically, they can be lumped into two groups: the dysergy load of the individual and general ignorance of synergetics at the present time.

Nevertheless, they can be used to some extent. Meanwhile, the value of synergetic sessions is made beautifully apparent. For in such sessions, the synergic mode turns *on*, and it is always a delicious experience.

There is one other situation in which it is relatively easy to use these tools. When two syngeneers get together and knowingly use them in their interactions the effect is electric. And with it comes a full realization of the tremendous potential available to humankind.

Humanity is one.

For the sake of sanity, remember: First is the event, the initial stimulus; second is the nervous impact of the event, via the senses; third is the emotional reaction based on the past experience of the individual; fourth comes the verbal reaction. Most individuals identify the first and fourth steps, and are not aware that the second and third exist.

——A.E. Van Vogt (1)

CHAPTER 9

IMPEDANCES

Every person has to contend with a *dysergy load*—the sum total of all forms of dysergy with which he or she is confronted. Each element of dysergy is like a traffic signal whose electric control is out of order, so that it shows red, yellow, and green all at once. A driver who encounters such a light, with cars crossing in front and horns honking behind, is in a dilemma. It's "Stop," "Go," and "Be careful" all at once.

But isolated instances of dysergy can be handled. The real problem is the cumulative effect of many elements of dysergy, in a variety of forms, from a number of different sources. This cumulative effect is enormous; and it drags everyone down.

As a result, every human mind is in a state of chronic *attenuation*. It is only fractionally aware and operates at only a fraction of its capacity.

One consequence of this is that most minds are so attenuated that they are unaware of the attenuation. They are aware of the dysergy, at least some of it, although, like an iceberg, most is below the surface of awareness. But attenuation diminishes perception, confuses or shuts down thinking, and paralyzes the will. Occasionally someone tries to do something about it, but all too soon he or she becomes entangled again in the web.

There is, however, another consequence that gives grounds for hope. Each of us is far better, far more capable than he realizes. The human mind, even in an attenuated state, is truly one of the marvels of the universe, capable of fantastic achievements. The problem is to tap this enormous power, to awaken the sleeping giant within us.

The solution to this problem, in basic terms, is to eliminate dysergy—or at least to reduce it to manageable proportions. To help accomplish this, it is useful to know the major sources of the dysergy load of the individual. To date, four major sources have been identified:

1. Personal impedances.
2. Impedances of others (with whom one has a relationship).
3. Twangles—chain reactions between two closely associated persons.
4. Group and social dysergy.

It is possible for an individual to eliminate his personal impedances, or at least to reduce them to a tolerable level. There are many ways to do this; the techniques presented in this book constitute one approach.* It is for the individual to decide which approach works best for him.

Personal impedances can, in some cases, be cleared by the individual working alone. Others may find it easier to accomplish this with a friend of like mind, each "coaching" the other alternatively. Where impedances are severe, it may be wiser to obtain professional help until the individual is able to work alone. In the last analysis, however, only the individual himself can clear his own impedances.

This is not the case with the impedances of others. Where association is close, such impedances (sometimes called impots— *imp*edances of *ot*hers) can contribute greatly to the total dysergy load of the individual. It is possible, however, for the individual to *neutralize* impots, i.e., to eliminate his own reactions to them and to substitute synergic responses. This does not clear the impot, of course; and the going is not as smooth as it could be; but neutralizing an impot does make things easier.

Twangles are an enormous waste of time. Since it takes two to twangle, the intelligent thing to do is to terminate the chain reaction. This is not always easy to do—there is always a temptation to "get in one last good lick"—but this only sustains the twangle. The wisest

*One of the saddest of the games played by people interested in personal development is the rivalry between "schools," each insisting that its Way is superior to all others.

policy is to avoid twangles where possible and end them as soon as you can.

Group and social dysergy are of major importance. Again, it is possible to neutralize such dysergy but the individual alone cannot clear it. However, there have been some exciting recent developments in Group Synergetics and Social Synergetics. These are discussed later. Participation in these developments can help reduce the dysergy load of the individual from this source.

The elimination of personal impedances is one action the individual can undertake that will reduce his dysergy load considerably and enable him to "go synergic" more readily and more often. In the remainder of this chapter we will describe the major types of personal impedance. Later, methods for systematically clearing these impedances will be presented.

We have found it convenient to classify impedances into three main types:

1. Chronic reactions
2. Protodynes
3. Self-invalidations

Chronic Reactions

Emotions have survival value. They are part of our evolutionary heritage. This is true even of the so-called negative emotions such as fear, anger, grief, hate, guilt, and greed. Usually, these emotions produce dysergy, but they provide an immediate reaction to a stress situation, which mobilizes both mind and body for action.

However, once the emergency has been handled, the emotion is no longer appropriate. A synergic individual quickly traverses to synergy and clears the dysergy produced by the emotional reaction.

Unfortunately, few people are yet synergic. The dysergy remains, in the form of a *chronic reaction*. Thereafter, whenever the individual encounters a similar situation, the chronic reaction surges to the fore. The individual more or less repeats what "worked" for him before, despite the fact that no two situations are ever identical.

In the course of his life, an individual accumulates a number of chronic reactions. Sometimes these are dormant, because a situation similar to the original incident is not encountered. Sometimes they are in more or less constant restimulation, either because stimuli similar to those of the original situation repeatedly occur or because an internal circuit has become established so that the individual restimulates himself.

The individual, it should be noted, is aware of the reaction. He knows what he is thinking and feeling, and what he feels impelled to do. He may even regret or feel ashamed of the reaction. But he is driven by it. When it turns on, he is in the Reactive Mode.

Chronic reactions may take an infinite variety of forms. Fortunately, they can be classified according to the determinants that *trigger* them. Previously discussed, they are repeated here for convenience of the reader:

SET and SEL points
GET and GEL points
Pleasure thwarts
Pain threats
Survival threats or losses
Value threats or losses

The determinants are like push buttons; whenever one is pushed, a chronic reaction occurs, which more or less runs its course. It may take a minute or even up to an hour or more.

Fortunately, chronic reactions are relatively easy to clear. All that is needed is self-honesty, the desire to clear them, and a technique for doing so. Such a technique (Analytical Procedure) is presented later.

Protodynes

A protodyne is a "primitive forcing pattern" in the Identic Mode. Unlike a reaction, it is entirely unconscious. It occurs in response to a stimulus and consists of a set of Identic Mode linkages and blockages.

Protodynes can perhaps best be understood by comparing them to the behavior of a person under hypnosis or acting under a posthypnotic suggestion. The hypnotist says, "You will feel an itching sensation in your right cheek. You will scratch it but then you will feel a burning sensation in your right hand. You will wave your hand about to cool it but then feel a tickling sensation in your throat. You will cough and then clap your hands together three times. You will do these things when you see me scratch my right ear. When I snap my fingers, you will awaken. You will remember nothing of this trance."

He snaps his fingers, and the subject opens his eyes, perhaps mumbling that he must have dozed off for a few minutes. The hypnotist scratches his right ear. The subject scratches his right cheek, and then exclaims, "Ow, that's hot!" and waves his hand about. Then he coughs and suddenly claps his hands three times. He is a bit puzzled by the latter action, but explains it away as "just an impulse."

A protodyne is like that—a sequence of linkages and blockages that runs its course in response to a stimulus. *And practically everyone has them!* They are relatively easy to demonstrate.

Protodynes are not ordinarily installed under hypnosis, of course. They are "installed" in other ways. A protodyne, for example, might first appear when a child does something that angers his father and is slapped so hard it makes his ears ring. The child is confused, hurt, and angry. But, fearful of further offending his father, he suppresses his anger. He tries to do something to please his father, but his father is busy and pays no attention. So he pretends to be sick, but that doesn't work either. Finally, he gives up and plays a fantasy game in which he is noble and unappreciated.

Later, as a man, he has a job with a boss who is displeased with his work.

Boing—he has a feeling of being no good, inferior.
Boing—he has a desire to be good, to drive himself.
Boing—he acts in accordance with a set pattern.
Boing—he suddenly feels tired, slightly nauseated.
Boing—he indulges in a Walter Mitty daydream.

The sequence of "boings" is the protodyne—in this case, a sequence of linkages. The individual, however, is unaware of the "boings." *He is also unaware of the stimulus*—the boss who reminds him of his father.* The thoughts, emotions, actions, etc. that he carries out happen, one after another, just as if he were carrying out a series of posthypnotic suggestions.

When protodynes are activated, they operate the individual like a puppet on strings, without his awareness or consent. Protodynes are very powerful. They can turn on perceptions, thoughts, emotions, body states, actions, images, recollections, etc. They can turn on fear for no apparent reason, or an outburst of anger that is entirely out of

---

*He is, of course, aware of the boss as a person, but not of the fact that boss is a stimulus for a protodyne. In other cases, the stimulus usually escapes conscious notice—it could be the sound of a passing car or a dripping faucet, the sight of a porch swing or an unlit candle, etc. The hypnotist scratching his ear was likewise unnoticed consciously.

proportion to the irritating stimulus. They can make a person sick or even cause fainting. They can turn on a headache or make a person feel blue or depressed or just tired. They can make a person wake up at night and be unable to sleep. They can make a person confused and unable to think clearly, with ideas going around in circles.

Protodynes are responsible for a great deal of unhappiness in the world. Silently, operating below the level of consciousness, they cause behavior that is clearly irrational to others, but not to the individual himself. Because few people understand protodynes or even know of their existence, the tendency is to blame the individual for his irrational conduct. They do not realize that it is not the *individual* who does these things, that he is simply being operated by a protodyne.

Naturally, no one likes to think that he is being operated by a protodyne. And since protodynes are unconscious, it is understandable that many individuals would deny their existence, at least in themselves. But the fact is that nearly everyone has them. (And the only reason I say "nearly" is to avoid making an extreme statement that I could not prove. The fact is that I have been able to demonstrate the existence of protodynes in everyone I have coached.)

When protodynes are unusually severe, they cause nervous breakdowns (neuroses) or psychotic breaks. Our mental institutions are full of people who are largely operated by protodynes. Unfortunately, the way some patients are treated (so-called psychosurgery, aversive therapy, electroshock treatments, etc.) makes one wonder about some of the people in charge of these institutions. However, it is very difficult to clear protodynes in patients with severe mental illness.

Fortunately, it is relatively easy to "key out" protodynes in people who are not seriously ill mentally. Techniques to accomplish this are described later in this book. By "key out" is meant that the protodyne is rendered ineffective. The individual no longer is under their maleficent influence.

Knowing of the widespread existence of protodynes, and realizing how powerful they are and how deeply they control us without our knowledge or consent, one would think that *everyone would want to get rid of the damned things as soon as possible!* If we lived in a rational civilization, a planetary program of clearing all protodynes would be given immediate attention and highest priority. Such a program could be accomplished in a decade, at a fraction of the cost now spent on world armaments. And once completed, the leaders of the world, and their peoples, would be free of the Identic Mode thinking that causes wars, and it would soon be clear that massive spending on armaments was stupid, and the people of the world would soon be free of an enormous burden. So a planetary clearing program would actually pay for itself a hundred times over.

Unfortunately, we do not live in a rational civilization. However, a beginning can be made, starting with ourselves. Thus getting rid of our protodynes (and other personal impedances) is a highly synergic act: it not only frees ourselves, with an increase in happiness, well-being, and ability levels; it also helps bring a planetary clearing program one step nearer.

Protodynes are discussed here because it is helpful to know about them, even though it is unnecessary to contact them directly in order to eliminate their influence. There are many different varieties, and the specific linkages and blockages will vary greatly from person to person. Experience has shown, however, that most people have seven major prodynes. These are identifiable as follows:

A Need protodyne
A Control protodyne
A Hate protodyne
A Fear protodyne
A Grief protodyne
A Pain protodyne
A Guilt protodyne

A technique for reducing these protodynes, without contacting them directly, is included in the Appendix. It is called Protodyne Reduction Procedure. Its use is discussed later.

Self-Invalidations

The most important impedances of any case are the Self-Invalidations.

From the moment of birth—and even before that—each human being is engaged in a process of creative evolution. Creative evolution has been eloquently described by Henri Bergson (2):

> The existence of which we are most assured and which we know best is unquestionably our own, for of every other object we have notions which may be considered external and superficial, whereas, of ourselves, our perception is internal and profound. What, then, do we find?...
>
> I find, first, that I pass from state to state. I am warm or cold, I am merry

or sad, I work or I do nothing, I look at what is around me or I think of something else. Sensations, feelings, volitions, ideas—such are the changes into which my existence is divided and which color it in turns. I change, then, without ceasing. But this is not saying enough. Change is far more radical than we are at first inclined to suppose.

For I speak of each of my states as if it formed a block and were a separate whole... Nevertheless, a slight effort of attention would reveal to me that there is no feeling, no idea, no volition which is not undergoing change every moment: if a mental state ceased to vary, its duration would cease to flow. Let us take the most stable of internal states, the visual perceptions of a motionless external object. The object may remain the same, I may look at it from the same side, at the same angle, in the same light; nevertheless, the vision I now have of it differs from that which I have just had, even if only because the one is an instant older than the other. My memory is there, which conveys something of the past into the present. My mental state, as it advances on the road of time, is continually swelling with the duration which it accumulates: it goes on increasing—rolling upon itself, as a snowball on the snow. Still more is this the case with states more deeply internal, such as sensations, feelings, desires, etc., which do not correspond, like a simple visual perception to an unvarying external object...

...Our duration is not merely one instant replacing another; if it were, there would never be anything but the present—no prolonging of the past into the actual, no evolution, no concrete duration. Duration is the continuous progress of the past which gnaws into the future and which swells as it advances...

Thus our personality shoots, grows and ripens without ceasing. Each of its moments is something new added to what was before... It is an original moment of a no less original history.

This passage beautifully describes the creative evolution of a human being as a whole. But there are domains of experience and action in which an individual encounters situations with which he cannot at that moment cope.

Whenever such an encounter happens, and the judgment of being "unable to cope" is made, *creative evolution turns off* as far as that domain is concerned. Such a turn-off is called a *Self-Invalidation.*

It is important to be as precise as we can about this, because on such precision depends the effectiveness of the trackers' work effort. So let us make some clear distinctions.

The judgment of "being unable to cope," first of all, is better described as "being unRAW to cope" (Ready-Able-Willing). Second, it is desirable to give such a judgment a name. We call it a *Prime Determinant.*

As soon as a Prime Determinant occurs, creative evolution in that domain turns off. This is the Self-Invalidation.

What happens next? The situation is basically what Bateson (3) has beautifully described as a "double bind." The individual is (or judges himself to be) unRAW to cope with the situation, and *yet he somehow must cope with it*. The Identic Mode takes over and produces a protodyne.

It is vitally important to note here: *the only thing that activates a protodyne—the only reason it is used—is that a Prime Determinant has occurred and a Self-Invalidation has followed.*

It follows that all that is necessary to deactivate or "key out" a protodyne is to *clear the Self-Invalidation [S.I.]*. Once the S.I. has been cleared, the protodyne is no longer needed, and it gradually fades away.

How is the Self-Invalidation cleared? The answer is basically simple, though not always easy to execute. A Self-Invalidation is cleared *by turning creative evolution on again.*

Actually, creative evolution is never fully turned off—only partly so, in a particular domain in which the individual is unRAW to act or to experience. On many other levels, in many other domains, creative evolution is continued, so that the individual, as a whole human being, is no longer exactly the same person. He has accumulated much new experience, acquired new skills, learned new knowledge. All this can be brought to bear, together with the creative evolution, which is even then continuously occurring.

Techniques for clearing S.I.s are described later.

There are various different types of S.I. They include the following:

1. An agreement with another person, group, organization, etc. that dysergically restricts or usurps one's freedom of action, thought, etc.
2. A dysergic judgment, "I am inadequate," relative to a domain of action that is vitally important to the individual.
3. A self-rejection—a rejection of a part of one's being.
4. An acceptance of complete control by some other person, group, organization, etc.
5. A "self-control"—an internal imposition of control to conform to some external standard or to please someone.
6. A VIP-bond—i.e., a dysergic acceptance of the Value-Interest-Perspective of some group, class, ideology, etc.

7. A confusion of identity—identifying "I" with one's name, body, family, or genetic heritage, or membership, status, and/or role in some group, organization, etc.
8. A Noble Burden—a dysergic acceptance of the dysergy load of another because of pity, love, or grief at the death of a loved one.

There may be others.

Another important way to classify S.I.s is into simple and complex. A simple S.I. is one that occurs in response to a Prime Determinant, with resultant activation of a protodyne. What has been described above is a simple S.I. However, a person with one or more protodynes in chronic or recurrent activation—or simply in the grip of a transient or chronic reaction—may also encounter a Prime Determinant. This results in a complex S.I., one complicated by the concomitant existence of other impedances.

Complex S.I.s are more difficult to clear. Fortunately, when a program of systematic clearance of S.I.s is embarked upon, the mind of the tracker awakens to the effort and aligns with it. The mind then begins a kind of sorting operation, presenting S.I.s in the order in which they may best be reduced and cleared.

The encouraging thing about S.I.s is that they are relatively easy to contact and are not difficult to clear, provided work effort is properly focused and aimed. Indeed, many people can clear them entirely without outside aid. Expecially noteworthy is that it is not necessary to probe deeply into the unconscious to attack protodynes directly. They simply fade away, like the proverbial old soldier.

An individual, then, can clear all his chronic reactions and S.I.s, and his protodynes will fade away. This enables him to turn on the synergic mode much more easily and to stay synergic for long periods of time, provided his overall dysergy load is not too high. True, it is always possible for him to encounter a new Prime Determinant and to experience a new S.I. But this can be cleared.

Techniques for clearing these impedances are described in the next chapter.

CHAPTER 10

CLEARING IMPEDANCES

This chapter describes techniques for eliminating personal impedances—chronic reactions, protodynes, and Self-Invalidations (S.I.s). These techniques have been extensively tested and they work well, provided they are used as described and in accordance with certain principles. These principles are codified in the Tracker's Guide and the Coach's Guide, described later.

There are two ways to approach a case. One is to take impedances more or less as they come, relying on the mind of the tracker to sort them out in the best order. This has the advantage of emphasizing the uniqueness of each individual. Its disadvantage is that work effort sometimes bogs down; the tracker makes some progress and then "plateaus out," usually because an S.I. has not been cleared or sufficiently reduced.

The second approach is comprehensive—to cover thoroughly and systematically all chronic reactions and S.I.s, reducing protodynes as needed to help clear the S.I.s. This has the advantage of thoroughness—if the tracker quits before he is through, he knows where he has left off and can come back later. Its disadvantage is that the order may not be optimal for a particular individual. Even here, however, the mind does manage to sort things out surprisingly well.

Of course, it is possible to combine both approaches, and this is probably the best. Sessions in a comprehensive approach can be interrupted from time to time with "free" sessions. It may also be helpful to intersperse sessions that are *not* directed primarily at impedances—creative sessions, learning sessions, or simply "sweeps" as previously described.

It is possible for an individual of average or higher intelligence and self-honesty to clear all his impedances working alone, using these techniques. Indeed, it cannot be too strongly emphasized that it is *the individual's own work effort* that clears the impedance. The technique is simply a *synergic guide,* which avoids waste effort and tells the tracker what to do next. And even when a friend acts as a coach, the *tracker* is still always in charge and *uses* the coach to help him.

Individual work, then, should be part of every tracker's Work Plan.

However, other forms of work can be used, and when it is feasible, use of these forms may facilitate work effort and shorten work time. These other work forms are:

1. Togetherwork—in which two or three persons respond in turn to the queries of Operation Traverse, Protodyne Reduction Procedure, Operation Milquetoast, or other work programs. (See Appendix)
2. Teamwork—in which two individuals alternate as tracker and coach.
3. Group Tracking—in which a small group (usually three to eight people) track together on a topic of general relevance.

It is also helpful for the tracker to have one or more Work Goals and a Work Plan. Essentially, this is encompassed in a Master Directive. The Work Goals will vary from person to person, and the tracker can change them at any time; but if he has them clearly formulated, and their value to him is clearly stated, he is likely to accomplish more in less time. The same applies to his Work Plan.

The Tracker's Guide

The Tracker's Guide is a set of practices, policies, and principles which, experience has shown, facilitates work effort and helps to solve problems that may arise. It is suggested that the tracker study this thoroughly and refer to it periodically. Here it is:

1. Assume basic responsibility for your own progress. (Progress occurs as the result of the work that you do; the technique simply guides your work effort along synergic channels, and tells you what to do next.)
2. Observe, analyze, and evaluate with total objective self-honesty. (No pretense, avoidance, self-blame, self-punishment, etc.)
3. Invest an honest work effort each time you use a tool. (Do not use a tool as a mechanical ritual.)
4. Use each tool with intelligent precision. (*Know* the tool and apply it with common sense.)
5. Use each tool with an Information Source orientation. (A closed mind cannot make progress.)
6. Be willing to re-examine and re-evaluate any belief, attitude, motive, value, etc. (You may decide not to change but you are open to change.)

7. Explore sidelines freely but always return to track. (It isn't necessary to use the tools to force your mind into a rigid groove.)
8. Use Information Source and the Tracker's Guide on difficulties that may occur.
9. Open each session with Take Charge Procedure (described below).
10. Close each session with a Session Review. (It is amazing how often *new material* emerges during a Session Review.)
11. Do not use techniques of other schools unless they are clearly synergic with synergetics. (You are the judge of this; there are many excellent techniques of other schools. However, mixing incompatible techniques and ideas usually leads to dysergy.)
12. Do not use synergetics as a therapy for symptoms. (Symptoms are usually "solutions" to basic problems—not very good solutions, but they work. Analyze the problem first, get a better solution, and the symptom will fade away.)
13. Establish and execute a definite work program.
14. Periodically review the Tracker's Guide as a means of improving your work effort.

Take Charge Procedure, referred to above, consists essentially of the first three policies of the Tracker's Guide.

1. Assume basic responsibility for your own progress.
2. Observe, analyze, and evaluate with total objective self-honesty.
3. Invest an honest work effort each time you use a tool.

Take Charge Procedure is shorter than the Tracker's Guide as a whole and helps to focus work effort sharply. It is used at the beginning of each session.

Analytical Procedure

Analytical Procedure is essentially tracking on impedances, using the Belief-Attitude-Motive Triangle.

The tracker begins by selecting an impedance to track on, by CEDA sequence. He then describes his BAMs related to that impedance.

As a rule, BAMs are of two types: rational BAMs, and reactive BAMs. The tracker should focus on the reactive BAMs if he wants to hit pay dirt quickly.

Beliefs are analyzed into *data* and *interpretations,* as already described. When this is done, the irrational character of the belief soon becomes apparent, and the tracker is able to replace the belief with a more rational one. However, this is usually on an intellectual plane, unless the motive side of the BAM Triangle is also examined.

Motives are traced down to one or more of the determinants. For convenience, they are repeated here:

Self-Esteem Threat or Loss Points (SET or SEL Points)
General Esteem Threat or Loss Points (GET or GEL Points)
Pleasure Thwarts
Pain Threats
Survival Threats or Losses
Value Threats

When a determinant has been reached, the tracker *replaces* the reaction with a *synergic response.* This takes an investment of time, energy, and thought, but it is always possible to devise such a response. When it is found, it is *far more effective* than the reaction, which usually only generates a chain reaction.

Another very useful technique in Analytical Procedure is the use of Information Source to exploit moments of attenuation. Here is how it works:

From time to time, in the course of tracking, the tracker may go "off track." This is signalled, for example, by the following:

1. The tracker starts to daydream.
2. The tracker is "seized" by an emotion such as anger, fear, or grief—etc.
3. Thinking slows down or stops entirely or becomes confused.
4. The tracker becomes bored or drowsy.
5. The tracker becomes reluctant to continue or feels an urge to quit.
6. The tracker encounters a BAM against thinking.
Etc.

Attenuation is a turning down or a turning off of the rational powers of the mind, accompanied by a narrowing of consciousness. When it occurs, *it is a highly reliable sign that something hot has been encountered.* By adopting an Information Source orientation—regarding the situation as an opportunity to learn something new—the tracker can exploit attenuation to great advantage. In other words, attenuation is a beautiful opportunity to grow.

Analytical Procedure is a very powerful technique for tracking down and eliminating chronic reactions. When it is done properly, the tracker often finds that the BAM he starts with *covers an underlying BAM*. This discovery of previously hidden BAMs is one of the more delightful features of Analytical Procedure.

Powerful as Analytical Procedure is, however, it is not in itself sufficient to eliminate protodynes or self-invalidations.

The Query Sequence

Nothing turns the mind on better than a question requiring thought to answer.

In ordinary conversation, a question is asked to obtain specific information: What time is it? Is it raining outside? Where are you from? etc. Similarly, in a test or examination, the student is expected to produce an answer that shows that he knows a particular part of a subject. The teacher already knows the answer.

However, there are questions that neither the asker nor the responder knows the answer to—at least, not immediately. In addition, there are questions to which different individuals will give different answers.

This kind of question we call a *query*. The purpose of a query is not to get specific information or to test knowledge, but simply to *turn the mind on*. It's very simple, and very effective.

Actually, a query does not have to be put in the form of a question, as long as it evokes a search effort.

A single query is very useful. Much more powerful, however, is the *query sequence*—a series of queries designed to evoke a search effort along a particular channel. A coach, for example, (as will be discussed later) may use the BAM Triangle as a guide to asking queries that lead the tracker quickly down to a determinant and to the substitution of a synergic response for a chronic reaction.

Query sequences can be organized into groups about a particular theme. A number of these groups have been developed, four of which are included in the Appendix. These are Operation Traverse, Protodyne Reduction Procedure, Operation Milquetoast, and Creative Procedure.

Operation Traverse is an excellent way to get started in Individual Work or Togetherwork. It can also be used by a coach in Teamwork, especially by someone just starting out in coaching. All the tracker has to do is read each query and respond. The *query sequence* guides him along a path designed to evoke an effective work effort, without the tracker having to know the technical details of the process.

Operation Traverse consists of twelve query sequences around such topics as Mother, Father, Siblings, Sex, Reality, etc. Each query sequence is designed to be completed in a single session of about one hour, although this varies. It isn't necessary to complete a query sequence in one session—this is for the tracker to decide.

Protodyne Reduction Procedure is designed to reduce the intensity of protodynes, without actually outlining them in detail. There are seven query sequences, aimed at the protodynes associated with Need, Control, Hate, Fear, Grief, Pain, and Guilt. Protodyne Reduction Procedure can be used alone, or in conjunction with Operation Milquetoast.

Operation Milquetoast is designed to locate S.I.s and to help the tracker reduce them. There are five query sequences aimed at Thwarts, Threats, Losses, Justice, and Social Invalidations. The actual clearing of S.I.s is accomplished in Creative Procedure, using a technique called Creative Tracking, described below.

Creative Tracking

The major target of individual work is the Self-Invalidation. S.I.s are cleared by a technique called Creative Tracking.

The basic sequence of Creative Tracking is the FEDA Sequence: Formulate, Experiment, Develop, Apply. Unlike the CEDA Sequence, the FEDA Sequence is a slow process, carried out over a period of two or three days or longer.

The Formulate phase is done in a single session. It consists of the following steps:

Step One: Define the Self-Invalidation. Describe it in words, as clearly as you can. Identify it as one of the major types—Agreement, Judgment of Inadequacy, Self-Rejection, Acceptance of Control, Self-Control, Value-Interest-Perspective Bond, Noble Burden, or Confusion of Identity. (If it doesn't fit one of these types, then you've discovered a new type!) Formulate a clear idea of the S.I.

Step Two: Analyze the Prime Determinant. Describe the two horns: (1) UnRAW horn—What is it that you are unReady, unAble, or unWilling to do or experience? Why? (2) Forcing horn—What is it that forces or constrains you to do or experience this thing? Why?

Step Three: Apply Dysergy Converters. These are described in detail below. Briefly, there are five of these:

Information Source
Mutation Source
Unique Focus
Buckmaster
Chinese Waterhammer

It is unnecessary to use all five of these; usually one or two will suffice. The purpose of the Dysergy Converters is to convert dysergy to synergy. Applied to the Prime Determinant and the Self-Invalidation, they introduce a new perspective and evoke creative evolution in these areas.

In using the Dysergy Converters, do not expect instant results. Creative evolution works slowly, sometimes taking twenty-four to forty-eight hours. Suddenly, at an unexpected moment, a B.I.* occurs. The important thing is to turn creative evolution on and let it take its course.

Step Four: Neutralize, Analyze, and Finesse Impots and Sociodynes. The Self-Invalidation is the key link in a causal chain that thwarts and handicaps the individual. A protodyne is a response to an S.I. and a Prime Determinant; a Prime Determinant, in turn, is produced by a sociodyne or an impot. These, in turn, have long causal chains behind them.

It is impossible for the individual, alone, to clear the impedance of another—only the other can do that. However, it is possible to eliminate one's *reaction* to an impot. Such a process is called *neutralizing* the impot. The impot remains, and it causes problems; but the individual is free to manage the total situation in a non-dysergic way.

In a similar way, it is possible to neutralize a sociodyne. This, too, frees the individual to a certain extent, even though the sociodyne continues to produce dysergy. With relevant impots and sociodynes neutralized, it becomes that much easier to manage the Prime Determinant and the S.I.

It is also possible to go further, by analyzing the impot or sociodyne, and devising a way to finesse it. By *analyzing* an impot is meant tracking it to a reactive BAM—especially to one or more

---

*B.I.—Burst of Insight.

determinants (such as a Self-Esteem Threat Point—see chapter 7). The analysis may not be wholly accurate, but it will provide a basis for understanding the impot. This, in turn, will make it easier to avoid reacting to it and will help the syngeneer to finesse it.

By *finessing* an impot is meant finding a way to achieve your goals without stimulating the impot and with the least possible interference from it. Basically, this is a matter of outsmarting it. Since the impot is reactive in mode, this is easy to do, *as long as the syngeneer is not himself reacting*. Neutralization should always be done first. If the impot is not neutralized, a twangle is likely to occur.

Neutralizing, analyzing, and finessing a sociodyne is similar in principle, though different in detail. A sociodyne is an Identic or Reactive Mode pattern embedded in the social matrix (see chapter 16). It operates on the individual through patterns-of-expected-conduct associated with his roles, statuses, and memberships in various groups. By focusing on his reaction to such patterns, the syngeneer can neutralize a sociodyne. Analysis involves identifying the various reactive BAMs that various parties (individuals and groups) trapped in the sociodyne tend to adopt. Finessing is accomplished by outsmarting it.

Step Five: Create a Synergic Ability. A Synergic Ability is an ability to operate in the synergic mode in a particular area. In simplest form, it consists of a set of mental operations (like a CEDA Sequence) that generates synergy when it is executed. By defining such a set, which will enable him to cope effectively with the Prime Determinant, the tracker creates a Synergic Ability.

Often it will be found that the tracker's mind has already worked out such an ability, but has not defined it clearly or put it into effect. The previous step of the Formulate phase will have turned on mental processes that sooner or later will reach fruition. Sometimes the tracker will find that, paradoxically, the synergic acceptance of a limitation—bringing Responsibility-Freedom-Power (RFP) into balance—is useful.

Two other processes may assist the formulation of the Synergic Ability. One of these is the Determination—the things the tracker does to get his own way in spite of everything. Although this is usually reactive in mode, it is relatively easy to traverse to synergy on it, converting the Determination into a Synergic Ability.

The other is the Compensation. Just as a person who becomes blind develops hyperacute hearing, so does a person with an S.I. in one area often develop a compensatory ability in another. This is the Compensation. Locating and tracking on the Compensation often leads to the development of a Synergic Ability in the turned-off area.

Creativity cannot be reduced to a formula; it takes an infinite variety of forms. Usually, there is a period of groping, of trial and error, of following false trails and going into blind alleys, before a solution flashes into awareness. It is important to realize this and to keep tracking. Sooner or later, the Synergic Ability will emerge.

When it does, the tracker is ready for the next phase:

Experiment. The tracker devises a situation in which he can apply the Synergic Ability and does so, observing what happens.

Develop. In this phase, the experience gained in the Experiment is examined and evaluated. The concrete experience often brings out new aspects not foreseen in the Formulate phase. In addition, creative evolution keeps on rollin' along, like the Mississippi; it cannot be wholly stopped, only temporarily diverted in its course. In the Develop phase, the Synergic Ability is clarified, modified, and in general improved.

Apply. In this phase, the tracker continues to experiment and develop the Synergic Ability until it becomes an integral part of his repertoire of skills. As he does so, any protodynes associated with the S.I. gradually fade away.

After a FEDA Sequence has been successfully completed, there is one more step the tracker may wish to take. This is to formulate a Rational Emergency Pattern for dealing with future Prime Determinants in case any are encountered.

Creative Tracking is a powerful tool, and in principle should work for everybody, because creative evolution is an integral part of human being. However, difficulties may be encountered by some, because protodynes are triggered which themselves block an effective work effort. When this happens, a session or two of Protodyne Reduction Procedure may be of value.

If this doesn't work, it simply means that the dysergy load of the individual is too great, and some form of psychotherapy may be indicated.

Dysergy Converters

There are five Dysergy Converters, mentioned above. In addition to their application to Prime Determinants and Self-Invalidations, they are very useful in any situation of high dysergy.

Information Source, already described, is the adoption of an attitude that the situation is an opportunity to learn, to discover something new, to view things from a different perspective, etc. The tracker focuses the attitude especially on the dysergy of the situation, which often manifests itself in "either-or" terms, confined to a single perspective. A characteristic result is the emergence of a new perspective with a shift to a "both-and" prospect. This is delightful when it happens.

Mutation Source is a generalization of Information Source to the "whole-being" level. Information Source is primarily conceptual. In Mutation Source, the tracker regards the situation as an opportunity to evolve, to experience a transformation of one's being. Often this takes the form of a giving up or a changing of some value previously held somewhat rigidly. Mutation Source is useful when the tracker encounters dysergy that makes him particularly uncomfortable or deeply discouraged. Regard this as a sign that a breakthrough is about to occur—and, after a short period of time, a deep insight often does emerge. Mutation Source is useful when Information Source alone seems to be insufficient.

Unique Focus is an effort to contact directly the process of creative evolution that is continuously going on in everyone and to focus it upon the dysergy of the situation. A human being is always in a state of change, of development, of growth, at a holistic level; but if he is gripped by a fixed, partial viewpoint based on past encounters (often an effort to prevent something unpleasant from happening again or to hold onto something pleasant in fear of its being lost), he is shut off from contact with this creative process. In Unique Focus, the tracker regards the situation *as a whole* as unprecedented and himself as a unique person at a unique and unprecedented moment in his life history. Whatever was once too terrible to do or experience is no longer so at this new and unique moment; what was once so compelling has lost some of its power. *There is a creative way out.* Sooner or later, if you search long enough, from the perspective of Unique Focus, you will find that way.

Buckmaster is named after the smart billygoat, who, after butting his head against a stone wall a few dozen times, decided to look for a gate through, a way around, or a way over, that wall.

The Chinese Waterhammer is a name conjured out of nowhere to signify a powerful technique for handling situations of overwhelming dysergy. The tracker coolly steps back from the perspective in which he is viewing the situation, takes a fresh look, and asks himself, What are the unique advantages of the situation? Such advantages are always there, because there are an infinite number of perspectives from which any situation can be viewed. Sooner or later, if the tracker keeps on searching, he will find a perspective that does contain advantages. Focusing on those advantages and utilizing them won't of itself clear the dysergy (if it is external), but it will make it more bearable and it may confound the dysergy sources.

These five Dysergy Converters are by no means the only ones; the reader may know of some of his own that are at least equally effective. We have found these five to be very powerful, however, and extremely useful in helping the tracker to transcend Prime Determinants.

One other very powerful technique deserves mention. By performing a phase shift and a prime shift on a Prime Determinant—or, indeed, any content—the tracker achieves a broad band perspective that is very effective in clearing dysergy. This technique is called *bracketing*. The order in which the operations are performed doesn't matter.

Conclusions

These, then, are some of the techniques currently available for clearing personal impedances—chronic reactions, protodynes, and Self-Invalidations. They work quite well when used by individuals who are ready, able, and willing to work according to the Tracker's Guide.

Not every person, of course, is ready, able, and willing to do so. For those who are ready and willing, but unable, psychotherapy is recommended. Therapy, in the hands of a competent professional, may enable the individual to reduce his impedances to the point where he will be able to work according to the Tracker's Guide.

As previously mentioned, it is helpful for the tracker to have one or more Work Goals and a Work Plan. What these are to be is for the tracker to decide. To illustrate a possible way to proceed, however, the following is suggested:

Work Goal: To eliminate all Self-Invalidations and stabilize in the synergic mode.

Work Plan: Solo, one session each day for about an hour, using the following:

1. Operation Traverse
2. Protodyne Reduction Procedure
3. Operation Milquetoast
4. Creative Procedure

Completion of this program will enable a person of at least average intelligence and self-honesty to stabilize in the synergic mode— provided he works systematically and thoroughly. See the Appendix for details.

CHAPTER 11

COACHING

Coaching is *not* psychotherapy. This needs emphasis because there are superficial similarities between the two. The main practical reason for this emphasis is that there are people whose dysergy load is so great that protodynes are in a relatively high state of activation. They are simply unable to track. Trying to coach them, therefore, is a waste of time. They need the help of a competent psychotherapist.

There are several major differences between coaching and psychotherapy.

1. The intent is basically different. In psychotherapy, the patient seeks aid because he has a personal problem he can't handle. In some cases, the patient produces so much dysergy that his family or the state can't handle it without special facilities.

In coaching, the tracker seeks assistance in clearing impedances because these are an obstacle to function in the synergic mode. He has a very definite, positive goal. If he did not have this goal, the impedances might still be a problem, but it would be a problem he could live with.

2. In most forms of psychotherapy,* the *therapist* is in charge. The patient expects the therapist to have special knowledge and skills that will be used to *treat* his condition. If the therapist is a psychiatrist, drugs may be used, as well as other forms of treatment. In terms of the Status Cross, the therapist is Super and Pro, the patient is Sub and sometimes Anti. In psychoanalysis and related forms of psychotherapy, a special relationship called a *transference* is formed, which is of basic importance in therapy.

In coaching, the *tracker* is in charge. He makes the major decisions; he evaluates the material produced; he is responsible for getting the job done. The coach is more than just an assistant; he is a partner in the enterprise; but he is always a junior partner. Frequently, coach and tracker change places; they form co-coaching teams.

---

*A major exception to this is Rogers's Client-Centered Therapy, which is probably the closest to synergetics of any of the various schools of psychotherapy.

3. In psychotherapy, the therapist is a professional who is paid a fee for his services. The fee may or may not be paid by the patient; but somebody pays, because that is how the therapist makes a living. Some psychotherapists believe that the payment of a fee for their services is actually a part of the psychotherapeutic process.

In coaching, no fee is charged. Often the coach gets coaching in return from the tracker. In other cases, the coach may be part of an organization such as a church—indeed, a minister in a church may, if he knows synergetics, do coaching as part of his ministry. Or the coach may be part of a commune or an economic cooperative, rendering this service because he is good at it. There are many ways, limited only by the creative imagination of human beings, in which arrangements can be made to enable a person to coach, without actually doing this on a "fee for service" basis. In the last analysis, people coach because they enjoy doing it, because it is a deeply rewarding experience. "Freely helping one another help himself" is as basic to coaching as payment of a fee is for psychotherapy. Fundamentally, coaching is an act of love.

What, then, is coaching?

Coaching may be defined as the art and science of evoking in a human being the following processes:

1. A maximum use of his rational mind to solve problems.
2. A maximum use of his evolutionary power to grow and develop along lines of his own choosing.

Two points are worth nothing about this definition:

1. The *essential freedom* of the tracker to solve problems in his own way and to choose the lines along which to develop. If the coach interferes with this freedom—say, by solving the problem himself or by trying to direct the tracker's development—then he isn't coaching.
2. The *basic responsibility* of the tracker for the progress he makes. If the tracker is unwilling to accept this responsibility— in particular, if he overtly or covertly seeks to transfer this responsibility to the coach (or to the technique used)—coaching becomes impossible.

The coaching situation might at first glance appear to limit the power of the coach to help the tracker. Actually, it amplifies this power. It insures maximum work effort by the tracker. It frees the coach from the burden of an impossible responsibility—trying to do for the tracker what only the tracker can do. And it provides the opportunity for the coach to use the tools of synergetics under the most favorable conditions.

The coaching situation might at first glance appear to offer little or nothing to the tracker since he apparently cannot expect the coach to solve problems the tracker has been unable to solve all his life. But the fact is that a coach *does* help, in many ways. His mere presence, as an empathetic (*not* sympathetic) listener, stimulates and encourages the tracker to "get things off his chest." The need to communicate in words in itself *turns on* the rational mind. One of the primary tasks of the coach is to observe for *moments of attenuation* (described in chapter 10). This is the main difficulty in tracking; by detecting attenuation when it occurs and bringing this directly or indirectly to the attention of the tracker, the coach helps greatly.

Through these things alone, a coach, serving as junior partner of the team, helps the tracker. However, as a coach gains in experience, and as he makes progress in his own case, he finds that he can greatly facilitate the tracker's work effort by *asking queries*.

The purpose of a query is *not* to gain information for the coach; nor is it to "test" the tracker's "knowledge" the way a teacher tests a student. The purpose of a query is *to evoke an effort of thought or of work* by the tracker. It is vitally important that the coach understand this. Often a coach sees a solution to a problem that puzzles the tracker, and the temptation to give advice is strong. The coach may successfully resist this temptation directly, but indirectly yield by asking "leading" questions designed to get the tracker to see the coach's answer. This, however, defeats the purpose of the query, which is to evoke an effort of thought and/or work. Also, an answer may be "right" from the coach's point of view but "wrong" or "inadequate" from the tracker's point of view. No matter how wise and clever a coach may be, he cannot really know the tracker's mind as well as the tracker does.

In using Operation Traverse and other sets of query sequences (see the Appendix), the coach automatically is asking queries. But a good coach will be able to *supplement* these by queries of his own, better adapted to the *specific* material and context communicated by the tracker. While this should not be overdone, an experienced and competent coach can help a tracker greatly in this way.

In asking queries, the seven chief interrogatives are useful: What? Why? How? How much ? When? Where? Who? Each query evokes a different kind of work effort.

"What?" evokes a work effort of formulation, of specification. It is of value in clarifying confusion and in evoking an objective attitude.

"Why?" evokes an effort to solve a problem, to account for something, to explain. Sometimes it evokes a rationalization, which is not desired, but which itself indicates a "need to defend" which the coach can then fire queries about.

"How?" evokes an effort to describe in detail, which is useful in clarifying and specifying those details.

"How much?" has the value of introducing degrees between extremes. It is useful when traverse from Reactive to Uniordinal Mode is sought. It also helps to generate perspective.

"When?" elicits recall. The recall of specific incidents of the past, describing the parts of the incident in the order in which they occurred, is useful both in "blowing steam" from such incidents and in providing the data upon which a present-time belief may be based.

"Where?" is occasionally of value in pinning down specifics of an incident.

"Who?" has value especially in reference to authoritarian and protective figures of the past. Such figures may play an important part in determining unconscious attitudes.

Although interrogatives are useful, a query does not actually have to take the form of a question. An incomplete or deliberately ambiguous statement can achieve the same effect—coaches soon become very clever at this. The important thing is that the query evoke an open-minded, honest search by the tracker, in an Information Source spirit.

In addition to the interrogatives, *query targets* are useful. The Belief-Attitude-Motive Triangle is an example of such a target. Without mentioning "belief," "attitude," or "motive," the coach can fire queries *aimed,* for example, at getting the tracker to express a belief, and then to examine the *data* and *interpretations* that led to this belief. Similarly, a coach can fire queries that lead the tracker through a Consider-Evaluate-Decide-Act (CEDA) Sequence, without the tracker being aware of the operations involved.

Coaches soon become so adept at firing queries that they tend to overdo it. If a tracker comes to *depend* on the coach for queries, he thereby yields his basic responsibility for his own progress to a degree. This is a dysergy trap that should be avoided. Most of the time a coach should simply be listening and watching for attenuation. *The tracker, not the coach, is in charge.*

The Coach's Guide

Like the Tracker's Guide, the Coach's Guide is a set of policies, practices, and principles that defines the art and science of coaching. When and if problems arise, application of the Coach's Guide is often helpful in solving these problems. The Guide also serves to distinguish coaching from psychoanalysis and other forms of psychotherapy, hypnotism, the relationship of guru to student, etc. It is possible for a person to call what he is doing "coaching," or even to believe that what he is doing is coaching, when actually it is not. The Coach's Guide defines the difference.

The Guide is stated below. It is suggested that the coach study this, mull over it, and review it regularly. Indeed, it is a good idea for the coach to memorize it and review it before each session. It should also be clearly understood that the tracker can at any time request the Coach to review the Guide, and to accede to a tracker's complaint that he has not been following the Guide. This protects both the tracker and the coach.

1. Coaching is the art and science of evoking thought and self-directed evolution.
2. The tracker, not the coach, is in charge.
3. Progress results from the work-effort of the tracker.
4. Progress is measured by the Mode Ladder.
5. The effectiveness of coaching depends on the degree of rapport that exists between coach and tracker.
6. The coach listens, unless there is a good reason to say something.
7. The coach observes for attenuation.
8. If attenuation occurs, the coach asks queries.
9. The coach does not evaluate, interpret, or give advice.
10. The coach does not give sympathy.
11. If emotional discharge occurs, the coach calmly lets it run its course.
12. The coach treats everything reported in a session as confidential.
13. The coach asks that each session begin with Take Charge Procedure (see previous chapter).
14. The coach asks that each session be ended with a Session Review.
15. If at all possible, the coach does not permit a session to end unless the tracker is reasonably objective and cheerful.

No doubt there are other policies and principles that could be added to the Guide; but these fifteen provide a sound basis for coaching.

Rapport

The dictionary defines rapport as a "relation of harmony, accord, conformity, affinity—especially in an intimate and harmonious relationship." In synergetics, rapport is used essentially in this sense, with certain precise qualifications. These are:

1. Rapport is determined by the degree of synergy, empathy, and communication that exists. This is symbolized by the Synergy-Empathy-Communication Triangle (SEC Triangle).
2. Rapport is optimum when the status of coach and tracker are equivalent on the Status Cross (see chapter 4).

Let us consider these qualifications more closely.

The SEC Triangle provides a basis for evaluating the degree of rapport that exists, and also for systematically improving rapport. Each leg of the triangle mutually reinforces the other legs, so that there is a synergic relationship. *Synergy* refers to those interactions between coach and tracker that promote both the coaching effort of the coach and the work effort of the tracker. *Empathy* refers to the mutual understanding the coach and tracker have for each other— each comprehending the other's viewpoint without necessarily agreeing fully or adopting it as his own. *Communication* refers to the effective, two-way interchange of data, ideas, etc. between coach and tracker.

Synergy promotes empathy and communication.

Empathy promotes communication and synergy.

Communication promotes synergy and empathy.

Rapport is clearly distinct from the relationship of *transference* that exists between analyst and patient in psychoanalysis. Fenichel (1) defines the transference as follows:

The patient misunderstands the present in terms of the past; and then, instead of remembering the past, he strives, without recognizing the nature of his action, to relive the past and to live it more satisfactorily than he did in his childhood. He "transfers" past attitudes to the present.

These attitudes are transferred to the analyst, who is regarded with mixed feelings of admiration and hostility. The analyst works through the "transference neurosis" that thus develops.

It is, of course, possible for a transference to develop between tracker and coach. However, coaching becomes impossible under these conditions. In fact, this rarely happens, because strict adherence to the Tracker's Guide and the Coach's Guide prevents it.

Rapport can reach remarkable heights. Indeed, the phenomenon of Totaltalk, discussed in chapter 5, was first observed in rapport between coach and tracker, when the synergic mode turned on in both. The effect is electric!—and indescribable in words.

Work Effort and the Mode Ladder

The Mode Ladder, discussed in chapter 2, is the basic "yardstick" by which progress is determined in synergetics. The coach will find it very useful in evaluating where the tracker is with respect to a particular impedance and the kind of work he must do to climb the next rung of the Mode Ladder. There are four major types of work effort.

1. Work of Contact. This is the work required to bring material from the Identic up to the Reactive Mode. Work effort is seemingly slow in this form, and often the tracker seems confused and uncertain. Focus on nonverbal aspects often facilitates work of contact.

Although the coach seldom focuses on material at the Identic Mode level, sometimes the tracker naturally moves into this level. All other things being equal, it is generally wise to assume that the tracker's mind as a whole "knows what it is doing"—that it is presenting what is most appropriate at that time. The coach cooperates with this effort, firing queries designed to promote contact with and awareness of nonverbal contents, helping the tracker to bring these into clear awareness.*

2. Work of Discrimination. This is the work done by the tracker in climbing from the Reactive Mode to the Uniordinal Mode. Emotional discharge not infrequently occurs here and should be permitted and, indeed, gently encouraged. However, sometimes the tracker becomes involved in a dramatization of his emotion—playing it over and over

---

*An exception to this would be a tracker who has previously undergone psychoanalysis and who has developed the habit of "free associating," reporting dreams, etc. A wise coach will avoid such practices and remind the tracker of the basic differences between coaching and analysis, reviewing the Tracker's Guide, etc.

again like a phonograph record. Discharge does not really occur unless the tracker invests a certain amount of objectivity and rational re-evaluation of the material.

Work of Discrimination is not all emotional discharge, however. Basically, it involves a move from a two-valued orientation—black or white, good or bad, pro or con, super or sub, etc.—to a *many-valued* orientation that recognizes degrees between extremes.

3. Work of Perspective. This is the work done by the tracker in climbing from the Uniordinal Mode to the Multiordinal Mode. Basically, it involves consideration of two or more perspectives, seeing things from several different points of view. Work of Perspective is largely rational and verbal, but it can and often should include a consideration of nonverbal aspects. Progress is often rapid in this form of work, at least more rapid than Work of Contact and Work of Discrimination.

4. Work of Synergy. As perspectives multiply, the tracker is sometimes confronted with apparent conflicts among the various perspectives; and he can readily find himself "spread too thin," aware of many different perspectives without a unifying theme tying them together.

Work of Synergy involves focus on the interactions between perspectives, with particular reference to those interactions that promote both while impeding neither. It also involves a search for a more basic perspective that encompasses all perspectives under consideration. Sometimes a period of groping, of trial and error, of mullwork, is involved. Then, suddenly, the sun breaks through the clouds. *Synergy emerges,* with an electrifying surge that is beautiful and exhilarating.

Coaching is a deeply rewarding activity. To share with another human being the sometimes painful but eventually mind-transforming experience of clearing his impedances and turning on the synergic mode is a sublime privilege. And with this sharing there emerges a renewed hope for the future of humankind, that the age-old dream of a brotherhood of man will soon become a living reality.

PART THREE

*Group
Synergetics*

CHAPTER 12

THE HUMAN STUDY GROUP

Group Synergetics is the art and science of generating synergy and reducing dysergy in small human groups.

There are so many varieties of human groups that it would not be feasible to discuss the application of synergetics to them all here. Instead, we will select one type of human group—a group of individuals organized for two specific purposes: (1) to help each member of the group to stabilize in the synergic mode; (2) to evoke the synergic mode of function in the group as a whole.

Such a group is called a Human Study Group.

There are three synergies about a Human Study Group that are worth noting. First, membership in a Human Study Group facilitates the work effort of the individual. The group is a source of coaches, for one thing. For another, the insights each gains can be shared with others for the benefit of all. Conversely, the synergic development of each member, accomplished in individual sessions, facilitates the synergic evolution of the group.

Second, when the synergic mode turns on in the group, a new whole emerges that is indescribably beautiful! There is love and understanding and teamwork in thought and action to a degree that is rarely achieved among human beings. A group united by a common danger or sharing in a high cause sometimes achieves a similar relationship, but it lacks the depth of mutual understanding, the *total empathy,* that may emerge in a Human Study Group.

Third, a group of stables united as a synergic team may undertake specific goals together with the prospect of high achievement—a prospect further enhanced by the fact that they can *knowingly use synergetics* in achieving these goals. The potential here is tremendous! Thus, a synergic team is basic to high enterprise.

Group Formation

A true group is a collection of individuals having a common goal or goals.

The goal is the basic cement that binds the group together. When any member no longer freely desires the group goal he is no longer a member of the group, though he may be physically present and even in a position of influence. There are many collections of individuals that are not "true groups" as defined above.

Two distinctions should be made at this point. The first is the distinction between a *group* goal and an *individual* goal. An individual goal is one sought by an individual for his personal benefit; he may, in pursuing it, take due regard for the individual goals and interests of others, but his goal is a personal one. A *group* goal, on the other hand, is one chosen by a group through a decision that each person freely shares or fully approves.* The reason this distinction is important is that people often confuse their group and individual goals and get all mixed up as a result.

The second distinction is between a *goal* and an *aim*. A goal, as defined here, is a desired future state that may be specifically achieved. Buying food for the next week is a goal; getting a high school diploma or a college degree is a goal; achieving stability in the synergic mode is a goal. An *aim,* as defined here, is a *general* purpose that can never be completely achieved though it may be partially fulfilled on a number of specific occasions. Generating synergy is an aim; learning mathematics is an aim; loving another is an aim. Again, confusion between goals and aims, as here defined, is often a source of dysergy.

Applying these distinctions to group formation, it is clear that the goal for a group should be a *group* goal, not an individual's goal for himself or *even his goal for the group*. He may *propose* a group goal but only when the group freely chooses it does it become truly a group goal.

It is also desirable that the group select a group *goal,* not an aim. There are at least two reasons for this. First, a goal is easier to understand and simpler to achieve. Second, an aim commits the group members to an indefinite future together, which is generally not a good idea. If the group achieves its goal, it can then freely choose another or disband as a group—so nothing is lost by choosing a goal rather than an aim.

In forming a Human Study Group, a first consideration is group size. Experience has shown that about three to eight people is

---

*This is a *synergic* group goal. Many groups have their goals chosen for them by a Leader; the members are expected to accept these goals, usually without question or consultation. Sometimes there are conflicts between two or more would-be Leaders of subgroups.

desirable, with five or six an optimum number. If a larger number are interested, they can meet together for some purposes, but should be divided into smaller groups for actual work together.

Some individual usually has to take responsibility for forming the group. He becomes, temporarily, the Group Monitor—a role he performs until the group decides upon a group goal or goals. He then relinquishes the role, although the group may ask him to assume it again. Once a group gets going, however, the role of Group Monitor should be rotated so that everyone gets a chance to do it.

In the first meeting or meetings, two preliminary objectives are usually desirable. The first, especially if some of the members are strangers or do not know each other well, is for each to introduce himself briefly, describe his work or activity, his background, his reasons for being interested in synergetics, etc.

The second is to gain some understanding of synergetics. There are various ways this might be done. If the Group Monitor has a good knowledge of synergetics, he can present its basic ideas in short (five to fifteen minutes) presentations, followed by a group discussion. Another approach is to have each member read up on one or more of the basic ideas, present it briefly, and have a group discussion. This process can be continued for several meetings.

Soon, however, the group should decide upon its group goal or goals. What these are will depend upon the group; they need not be those explicitly described here, or they may include these goals with others added.

Goal selection should be by *unanimous agreement.* "Majority rule" is not necessary in a small group, and unanimity is necessary if all are *freely* to share in the group goal or goals. If a person dislikes a goal strongly desired by the other members, the wisest course is for him amicably to drop out of the group. This can be done without disturbing friendly relations.

"Free adoption of group goals" is like a gate. When a person freely adopts group goals he goes in the gate. When he no longer freely holds them, he goes out again. In general, the people "in" a group are usually not in it 100% of the time, but pass in and out of the gate from time to time.

Types of Groups

It is convenient to divide groups into types, since if a syngeneer can recognize the type of group he is in, he can deal with a situation more effectively.

A *mixed* group is one whose members do not all have the same group goals. Some may be there out of friendship or some other reason; or there may not be realization that the group goal is not held in common; or the group goal may not be clearly understood by all concerned. Most human groups are mixed groups. Such groups are useful and no negative judgment is intended here.

A *true* group is one whose members all freely adopt one or more clearly formulated group goals. A true group is the basis for all higher types of group. It always has a higher spirit and greater enthusiasm than a mixed group.

It should be noted that a collection of individuals may be a mixed group in some situations and a true group in others.

A *cleared group* is a true group in which all major clashes of personality, methods, etc. have been eliminated. Such clashes are called *twangles*. A true group usually submerges its twangles at first, in the interest of the group goal; and it may achieve the goal without clearing the twangles. If problems or obstacles are encountered, however, twangles tend to emerge and disturb group function.

Various special methods have been developed to eliminate twangles. These are described later.

A *synergic team* is a cleared group operating in the synergic mode as a group. It is not necessary that all members of the group be stables, but it helps! Very few synergic teams have appeared in history, and their advent in large numbers would constitute a major turning point in human affairs.

In chapter 5, a number of synergic team functions were described—Affinity Make, Empathy Make, Semantic Telepathy, Synapse, Franktalk, and Totaltalk. The knowing and habitual use of these tools by members of a cleared group catalyzes its transformation into a synergic team. A new, synergic whole emerges that is greater than the mere sum of its parts. The effect is electric and cannot be adequately described because common referents do not exist in ordinary groups.

A Program for a Human Study Group

In the remainder of this chapter, a sample program for a newly organized Human Study Group will be outlined. This is presented only as an illustrative guide, not as a prescribed procedure. Each group freely chooses its own goals and adopts its own approach. For convenience, the program is divided into four phases; however, there may be overlap and interactions among the various phases.

Phase One—Organization. In this phase, one or a few individuals take responsibility for organizing a Human Study Group. Notice of an initial meeting is given by bulletin boards, announcements at meetings, mail, telephone, personal contact, etc.

In the initial meeting (assuming the group is relatively small), the Temporary Group Monitor takes charge. It will be assumed that he has read this book and has extra copies available for others. He thus has some knowledge of synergetics, although he need not be an "expert" on the subject.

The meeting may begin by a short (less than fifteen-minute) statement of the purpose of the meeting and a brief presentation of a few basic ideas—no more than two or three. For example, the Monitor may present a definition of synergy and of synergetics as the science of synergy. He might then explain why synergy is important—first, because it improves the effectiveness of a complex system, and second because when enough synergy is produced, a new whole emerges that is greater than the sum of its parts—called the synergic mode of function. He might follow this by some examples of synergy, such as think-feel synergy and the synergy between an individual and a group of which he is a part. He might close by briefly describing his own background and interest in synergetics and stating that he is serving as a Temporary Group Monitor until the group is organized and a permanent one is chosen.

The initial presentation should be *short* and followed by a period of group discussion. This can be started by a round robin, in which each person introduces himself and describes briefly his background and interests. The Monitor may then ask for questions and comments about his initial presentation.

In the discussion, the Monitor should do what he can to promote participation by all and domination by none (including especially himself). This can be helped by keeping more or less to a round robin format to give each a chance to express himself without, however, being forced to do so. To get things started, the Monitor may ask a general question, like asking each to give examples of synergy from his own experience. Once discussion is underway it tends to be self-sustaining, for awhile at least.

After the discussion has gone on for awhile, the Monitor should bring the group to a consideration of whether or not it wishes to continue as a group for awhile to study synergetics. He may briefly present two goals for consideration: enabling each person to increase his personal synergy, and promoting synergy in the group itself. He might then outline the phases proposed here, at the same time recommending that all interested begin reading this book; if they have already read it, to study it.

If the group (or some persons in it) decides to adopt these goals, at least tentatively, the Group Monitor then closes the meeting by a group decision on time and place for the next meeting. The individuals may stay around afterwards and socialize; this should be encouraged but not forced.

Phase Two—Education. Phase One may last more than one meeting, or there may be some overlap in the transition to Phase Two. Let us assume, however, a rapid transition. The group has decided to become a Human Study Group, each member freely adopting the two goals: (1) to help each member of the group to stabilize in the synergic mode, and (2) to evoke the synergic mode of function in the group as a whole. The group has also elected a Group Monitor, and each member has obtained and has agreed to read and to study this book. It will also be assumed that the group is meeting regularly—once or twice a week.

The Group Monitor proposes, and the group discusses, revises, and adopts a schedule for mutual education on various synergetic topics. Each meeting considers a particular topic. A particular member of the group is selected as discussion leader for each topic so that each member of the group gets a topic, in rotation.

A sample schedule might be the following:

1. The Mode Ladder (chapter 2)
2. The Human Study Group (chapter 12)
3. The Status Cross and SEC Triangle (chapter 4)
4. Tracking—the CEDA Sequence (chapter 7)
5. Tracking—the BAM Triangle (chapter 7)
6. Group Tracking (chapter 8)
7. Impedances (chapter 9)
8. The Synergetic Session (chapter 6)

The group may prefer a different schedule or may decide to modify the schedule at any time. The important thing is to have a definite schedule, with a different member each time as discussion leader, and with each member reading beforehand all relevant material on the designated topic.

Group discussion should be by short (fifteen-minute) presentations by the discussion leader, followed by round robin discussions by the group members. Each member should try to think of particular examples from his own experience to illustrate some point and to think of questions that need clarification.

The meeting should conclude by a brief review of the high points of the discussion by the discussion leader. During this review any member may interrupt to clarify or evaluate or illustrate some particular point.

The basic objective of the Education Phase is to help each member of the group achieve a basic understanding of synergetics, an understanding sufficient to enable him to start work on his own case and to enable the group to begin group tracking. Again, there may be some overlap in the transition to Phase Three. Desirably, Phase Two should be kept as short as possible; some groups may be able to omit it entirely.

Phase Three—Work. In this phase, each member of the group begins work on his own case, and the group starts group tracking. Occasionally, the group may decide to have another education meeting, but the main emphasis is on clearing impedances, preventing and clearing twangles, and learning to turn on the synergic mode.

Again, how the group proceeds is for the group to determine. What follows is a guide to illustrate one possible route.

The group breaks up into teams of two or at most three. The meeting is divided into two parts: one for individual work in synergetic sessions, and later a group tracking session. About twice as much time (or more) should be devoted to individual sessions as to group tracking.

In the individual sessions, one way to proceed is to go through the query sequences of Stabilization Procedure, as described in the Appendix. This can be done in either of two ways:

1. First, one goes through a session with the other, as coach, asking queries. Then the two change places.
2. Both go through a session together, each responding in turn to each query.

Be sure to follow the Tracker's Guide.

In the group tracking sessions, the group selects, by CEDA Sequence, a topic to track upon. Each member in turn then states his BAM about the topic. From time to time, one member may fire a query at another—designed to evoke thought, not to obtain a preconceived answer.

Any topic considered useful may be chosen. Some that have proved valuable in the past are:

Hiding	Sex
Responsibility	Freedom
Hate	Self-pity
Love	Blame
Selfishness	Regret

Periodically, instead of a regular group tracking session, the group should devote a session to twangles, as discussed in the next chapter. The objective is to transform the group from a true group to a cleared group. These sessions should be repeated until all twangles have been eliminated.

As Phase Three continues, sooner or later a member of the group will experience the primeburst—the first turning on of the synergic mode. This is a moment of joy and excitement, both in itself and because of its tremendous significance. Others will experience their primebursts, and members of the group will find the synergic mode turning on more and more often, and staying on longer each time. Meanwhile, group mode will steadily rise—from time to time group overdrive will turn on, as the group operates in the synergic mode as a group. The transition to Phase Four will have begun.

Phase Four—Synergic Team. The time will come when one member of the group will report that he has stabilized in the synergic mode. By this time, the group will have achieved such a depth of love and mutual understanding that the stable will know he can open himself completely to the group. Such an opening has been called "Running the Gauntlet": the group tests him—lovingly but stringently—firing tough query after tough query at him. No one passes judgment; only the individual himself can judge whether he is *fully* synergic. If it is found that he has made a mistake it is not considered a setback, but an opportunity to evolve.

Meanwhile, members of the group will be more and more engaging in that remarkable process called Joining Minds in Totaltalk. Each "reads" the other and "sends" to the other on all tracks and tempos, with a rapidity and scope that is mind-boggling. Above all, there emerges and blossoms a deep and beautiful love for one another—a caring/sharing that is so full and complete that a new entity emerges. This is the fabulous Empathic Communion of a synergic team.

One might expect this new entity to absorb the individuals who compose it. Yet, paradoxically, this does not happen. On the contrary, for participants in an Empathic Communion the unique individuality of each is not only preserved, it is enhanced and cherished!

A synergic team is finite and there must be limits on what it can accomplish. But these limits are so far beyond those of ordinary human groups that they seem almost infinite by comparison. The individual human mind is apparently rimmed by a set of concentric barriers that on the one hand limit consciousness and ability and on the other hand constitute thresholds to ever-higher modes of consciousness and action. To penetrate these barriers—to rise above these thresholds—is exceedingly difficult for the unaided individual, even for a stable. For participants in the emergent whole that is a synergic team, these barriers are more easily penetrated. This leads to yet another paradox: when a true group evolves into a synergic team, instead of stopping, its growth and development accelerates.

The synergic mode is so full of paradoxes and surprises that it has sometimes been called the paradoxical mode! Here is one more. One might expect a synergic team to be so fascinated by the beautiful new worlds of consciousness and activity open to it that the team would focus its time, effort, and thought entirely toward exploring these new dimensions. But this does not happen. The exploration goes on; but the team does not forget the humanity from which it sprang. Quietly, without fanfare, in ways that are invisible to the casual observer, synergic teams are engaging in High Enterprises—activities designed to promote synergy and reduce dysergy for their fellow human beings.

The world is so full of dysergy that it will be years, even decades, before these High Enterprises bear fruit. And there is still no assurance that humankind will not destroy itself before this happens, which is all the more reason for more and more people becoming synergetic stables as soon as possible and all the more reason for true groups to form and to evolve into synergic teams.

This is the cause—the High Enterprise—of synergetics.

CHAPTER 13

GROUP TRACKING

When human beings get together for purposes of communication, there are various ways in which it can be done.

One way is for one person to take the center of the stage and talk or lecture to the group. Theoretically, the speaker is giving data or knowledge about some particular topic to the group. This works for short periods of time; but unless the speaker is very, very clever, amusing, and stimulating, he loses more and more of his audience as time goes on, as students well know.

This is not Group Tracking.

Another way is by group discussion, each member of the group presenting his ideas and comparing them with those of others. Usually there are differences of opinion, whereupon each individual feels called upon to defend the "rightness" of his own ideas. Some do this by restating their views, going to great lengths to make clear the obvious, resorting to logic or appeal to authority to prove they are right. Others respond in kind; and soon an interesting phenomenon develops. It is as if each person splits into two parts—a real part and another made of straw. Alice talks to George's straw part and the real George talks back to the straw Alice. Each proves he is right from his viewpoint, but to an objective observer it seems they are no longer talking about the same thing. (And obviously to the "objective" observer, both are wrong, because *this* is the way it is, etc.)

Others "defend" their views by simply withdrawing into the privacy of their own minds, where, in secret, they *know* they are right (but there's no use trying to show Alice or George, they are so bullheaded!).

This is not Group Tracking either.

We emphasize this because there is a natural tendency for groups to proceed by group discussion or one-man shows, and these tendencies must be checked if Group Tracking is to be done effectively.

We have said what Group Tracking is *not*. What, then, is it?

Group Tracking is simply tracking by a group on a topic freely chosen by the group, in which each person examines his beliefs, attitudes, and motives about the topic for the purpose of understanding them and clarifying or changing them.

147

Any of the ideas and tools of synergetics may be applied where appropriate. Of particular importance, however, are the following:

1. The CEDA Sequence
2. The BAM Triangle
3. Information Source
4. Equivalence of Status on the Status Cross
5. The Synergy-Empathy-Communication (SEC) Triangle

It is done essentially as follows:

First, some member of the group is selected as Group Monitor. He participates in the group tracking, but has a special responsibility to watch to see whether or not the group stays on track. He does not "lead" the group. He does not coach the group or give "group therapy." He simply watches as he participates for various signs that the group is going off-track, such as:

1. One person starts to dominate the group.
2. The group stops tracking and goes into group discussion.
3. The group strays from the topic (unless this is done by group CEDA).
4. One person "withdraws" from the group—ceases to participate.
5. A "twangle" develops or threatens to erupt.

Whenever, in his judgment, the Group Monitor feels the group is off-track or about to go off-track, he interrupts and raises the question, "Are we on track?" This immediately takes precedence over whatever is going on. Each member of the group, in turn, gives his evaluation, and the group then goes through a CEDA Sequence. It may decide to switch topics or to return to the original topic; or it may decide the Monitor was in error.

A Group Tracking session is begun by selecting a topic by a CEDA Sequence. Each member in turn suggests a topic, a brief evaluation is given as each desires, and then a topic is selected by simple majority vote.

The group then tracks on the topic chosen. The simplest way to do this is by round robin. Each member examines, searchingly, his beliefs, attitudes, and motives about the topic. When his turn comes, he describes these as objectively and honestly as he can.

But more important than the description of these BAMs *is the search each makes into why he holds them.* This search is conducted in the spirit of open-minded, rigorously honest pursuit of the truth, no matter where this search may lead. If the search is not conducted in this spirit, nothing is gained. The individual is not really tracking, as this is defined in synergetics.

Also vital in Group Tracking is the attitude of others in the group. This attitude should be one of friendly caring, with a deliberate empathy make—each trying to understand and appreciate the other's viewpoint, without necessarily agreeing. If all members of the group knowingly adopt this attitude, an atmosphere is generated that is highly favorable to each person's search for truth. The defenses each of us uses in ordinary social intercourse are barriers to this search; the synergic atmosphere of loving concern adopted by all group members eliminates the *need* for these defenses and greatly aids the search.

Conversely, the rigorous self-honesty that one person uses in this search encourages others to do likewise. Once again, we see the wonderful process of synergy at work.

Each person will develop his own search paradigm; what is presented here is one way, not the only way. It works and may be useful as a guide.

Beliefs generally are of two kinds: rational and reactive. Rational beliefs are based on facts, ideas, logic, common sense; they are readily changed by the individual if he discovers a contradiction, an error, a better idea, etc. They are not charged with emotion. Reactive beliefs are heavily charged with emotion, and sometimes this emotion must be discharged before the beliefs can be honestly examined. Clearly, the examination of reactive beliefs is more productive.

In looking at a reactive belief, it helps to ask the questions: What data have led to this belief? How valid are these data? What interpretations have I made of the data? Are these interpretations fully warranted?

This begins the search. Sooner or later, however, the reason a reactive belief is held may be traced to some motive for holding it. And this motive almost always is found to be one of the determinants, discussed in chapter 7.

The reactive belief is then cleared, basically, by substituting a *synergic response* for the reaction to the determinant.

For example: Suppose the group has decided to track on the topic of hiding. One member might track as follows:

"What belief do I have about hiding? Well, it is a way of escaping from or avoiding a threat.

"Why do I adopt this belief? What data is it based on? Well, as a child, I chopped down some small trees with a hatchet when I was told not to, but I hid this fact from my parents. They didn't find out, so it worked. So I hide, or try to hide things when I fear disapproval, rejection, punishment, etc.

"Aha! I've already traced this belief to a motive—a General Esteem Threat point. And I also see that hiding may often greatly restrict my freedom of action—and if I am found out, it may put me in a worse light than if I didn't hide it. Synergic response: In some situations, hiding may be a good idea, but in most it is wiser in the long run not to hide."

The attitude side of the BAM Triangle is based on beliefs and motives. To fully clear the BAM, the tracker should look at this attitude. He asks himself, "What do I expect to happen when I adopt this belief and motive? How do I *anticipate*—prepare for—what I expect to happen?"

An attitude is nonverbal and sometimes difficult to describe in words. It manifests itself in a *sensory* and a *motor* set. In the sensory aspect, an attitude "tunes" the senses in such a way that they select certain data and block other data from reaching consciousness. Thus, in a hiding BAM, an attitude may lead a person to interpret an act of another as indicating he suspects something, when actually he does not. The hider may even respond reflexively in such a way as to call attention to what he is hiding!

In the motor aspect, an attitude sets the muscles in the body, tensing some and relaxing others. In a hiding BAM, it may produce a tendency to look downward or to avoid another's gaze.

In examining an attitude, the tracker focuses on the nonverbal aspect of his being. One way to do this is to look at the way his sense organs are *set* in terms of the belief/motive. Vision, hearing, smell, taste, and body sensations can be silently and systematically examined. Then the tracker may look at the way his muscles are set to act, in terms of what he expects may happen and how he prepares himself for—anticipates—what may happen.

The group may choose to have a period for an Attitude Scan—a period of silence after each tracker has found his belief/motive—in which he silently and nonverbally scans his attitude. This can be done in the way described above, but it doesn't have to be.

This, then, is Group Tracking—the open-minded, rigorously honest examination by each member of his BAM about a topic, in a synergic group atmosphere, for the purpose of understanding the

BAM and synergizing it. There are a number of additional ideas and techniques we have found useful. These will now be described.

Definitions and Examples

It often helps at the outset for each person to define as clearly as he can what he means by the group topic. This serves two purposes: first, it helps the individual to clarify in his own mind what he is tracking on, and second, it usually reveals that members have different concepts of the chosen topic. If "responsibility" is the topic, Fred's concept of responsibility may differ from Mary's. This doesn't mean that Fred is "right" and Mary is "wrong" or vice versa; it simply means they have different concepts denoted by the same word. If these distinctions are brought out at the start, mutual understanding is improved.

Further clarification and improved mutual understanding may be achieved by giving specific examples of what is meant. These can be imaginary or based on recollections of real incidents. Not infrequently, when Fred tries to give a specific example of what he means by responsibility, he finds that the definition he gave is incomplete. Specific examples also help each member to formulate his belief about the topic in concrete terms.

Queries

As people get experience with Group Tracking, their desire to help each other or to express an insight while it's hot leads to an interruption of the round robin format. This is acceptable as long as it does not get out of hand; the Group Monitor notes what is happening and continuously evaluates whether or not the group is really getting off-track, as described above.

One way in particular that this "cross-talk" may be helpful is the use of queries by other members to help one tracker in his search-effort. These queries are like the queries of a coach and should be used according to the Coach's Guide. As long as they do not distract the tracker from his search-effort, occasional queries can be very useful. However, it is all too easy for a competent coach to begin to dominate the situation by firing one query after another at the tracker. These may be valuable for the tracker, but they do not help the group. If this happens, the Group Monitor should intervene with the suggestion that the tracker pursue the matter further in a coaching session.

Twangles

A "twangle" is a chain reaction between two or more members of a group—Edith reacts to an action or statement by Bob, Bob reacts to Edith's reaction, Edith again reacts, etc.

Twangles sooner or later tend to emerge in any true group. A natural tendency is for one or both parties to suppress the twangle in the interests of the group goal. As a temporary measure, this may be desirable, but in the long run this leads to an accumulation of covert twangles.

It is recommended, therefore, that a true group make a deliberate effort to bring out and clear its twangles—to transform itself, in other words, into a cleared group. There are various ways of doing this. Here are two techniques that have been found useful.

1. Operation Whizbang. Each member of the group, in turn, says, "I don't like you, because..." to some other person in the group, stating a feature of the other person that he objects to such as the way he keeps interrupting or talks too much or doesn't take part, etc.

Then he is required to state why he permits this particular trait to bother him. In doing so, he is accepting responsibility *for the effect on him* of the trait. In other words, why would so calm and rational and admirable a person as he is permit such a thing to annoy him? This is the "Whiz" part of the operation.

The receiver of these remarks then regards this as an Information Source on his own case. He is not required to accept the criticism, which of course may not be accurate; nor is he expected to defend himself. Rather, he uses it as an opportunity to make progress. This is the "Bang" part of the operation.

Operation Whizbang works surprisingly well once it is understood. It lets off steam in an acceptable way and promotes group empathy. And it does help clear most twangles. However, it should not be tried until members of the group have tracked together for awhile, and have learned to know and trust each other. And it shouldn't be used on severe, chronic twangles.

2. Walking the Plank. This carries Operation Whizbang one step further, and should preferably be done only after a round or so of Whizbangs. A member of the group leaves the room. The other members then "exchange notes" on various ways he is avoiding, hiding, etc. in getting work done on his own case. They then call him in and give him the works.

Desirably, each member of the group should be required to Walk the Plank if anyone does. This preserves equality of status and prevents twangles from emerging.

With most groups, periodic use of Operation Whizbang and Walking the Plank will be very effective in clearing and preventing twangles; ultimately, all twangles will be eliminated and the group will become a cleared group. Unfortunately, situations sometimes arise that cannot be handled by these techniques.

Supertwangles and Twanglemakers

It sometimes happens that two individuals in a group react so strongly to each other that the very existence of the group as such is threatened. Such supertwangles are very difficult to handle and may well be too strong for the collective skills of the group to handle. This must be frankly recognized.

There are two options available. The first, which probably should be at least tried in most cases, is to try to resolve the supertwangle.

The group goes into a special state called Red Alert. This state desirably should be discussed at an early phase of group development so that everyone knows about it beforehand. Any member of the group (not just the Group Monitor) may call a Red Alert.

The group immediately interrupts its tracking to focus on the problem of the supertwangle. The immediate goal is to terminate the chain reaction as soon as possible. This may best be achieved by separating the twanglers, by dividing the group into two subgroups, each going into a separate room. Each subgroup can then coach the twangler for awhile, mostly to permit emotional discharge to occur.

If sufficient emotional discharge occurs, the twangler can then be asked to do a Switcheroo. Let us say that George is the twangler who is to undertake a Switcheroo and Howard is the other twangler (now in the other subgroup). George is asked to assume that he is Howard, holding an admittedly reactive BAM (as perceived by George—this is accepted as valid by the subgroup). George, pretending to be Howard, then tracks on the BAM, tracing it down to a (presumed) determinant.

By the time George goes through this, he may gain sufficient empathy for Howard that it will be possible for them to interact amicably again.

If Red Alert doesn't work, the wise thing to do may be to dissolve the group. Later, after a time lapse, one or more new groups may form. Usually by the time a supertwangle erupts a group will have formed co-coaching teams anyhow, so that work may continue on an individual basis.

A twanglemaker is an individual who pays lip service to the group goal and goes through the motions of Group Tracking, but whose *actions* either ignore or conflict with the group goal, and who in reality does not track. He may be so wrapped up in his own problems that he drags the group constantly into considering them to the virtual exclusion of all else. Repeated efforts by the Group Monitor to get the group back on track roll off him like water off a duck's back. Or he may be so full of hostility and rebellion that he constantly invalidates others in the group or synergetics itself. Or he may be a drifter—a person who drifts from school to school, dabbling in each so he gets an intellectual grasp of its ideas and techniques, but never seriously uses them. Drifting into a synergetic group, he uses it as the occasion for trying to introduce techniques of other schools, not in a manner synergic to synergetics, but in such a way as to prevent Group Tracking from occurring.

Twanglemakers operate according to the policy, "Help me to stay the way I am." They seek to avoid help by appearing to seek it. Once a twanglemaker is admitted to a group, he acts as a permanent dysergy source for the group.

Unfortunately, there is only one way to deal with a twanglemaker, and that is to dissolve the group and re-form it without him. This can be done by holding a special meeting of the group without him, to determine whether or not he is a twanglemaker. If the group decides he is one, it may continue meeting as a mixed group for awhile, permitting the twanglemaker to participate, while meeting as a true group elsewhere without him. There are various ways this may be tactfully done; but it is unfair to others in a group to be permanently entwangled.

An Example of Group Tracking

Suppose a group consists of three persons—A, B, and C. C is Group Monitor. They start a session by CEDA Sequence.

A: What shall we track on?

B: I'd like to suggest "hiding" as a topic.

C: I'd like to suggest "responsibility."

A: "Hiding" sounds like a good idea for getting better acquainted. "Responsibility" is something I'd like to straighten my ideas out on.

B: Actually, I can think of ways "responsibility" ties in with "hiding."

C: Me, too. I vote for "hiding."

A: "Hiding" it is, then.

B: O.K. What is "hiding"?

A: Maybe the question might be better put, "What are our beliefs about hiding?"

B: O.K. Hiding, to me, is doing something to keep somebody else from knowing about a thing you know about.

C: Ouch! That's a bit involved. I'd say I agree with it, but I'd like to take it apart. First of all, according to this belief about hiding, there has to be a person who is doing the hiding. Second, the hider has to know something. Third, there has to be a person or group who doesn't know that thing. Fourth, there has to be an action by the hider to prevent that person or group from finding out.

A: Say, that's pretty good. Then would you say, if any of these four things are absent, hiding can't be done?

B: You could go through the motions.

C: You'd have to *believe* all four things were present! Now that's interesting. What data would lead a person to this belief?

A: One thing might be his *motive* to hide something. We often believe a thing because we want to believe it is so, regardless of how true it is objectively.

B: A lot of times the person who hides may not know there actually is a person he is hiding it from. He's just playing it safe.

C: Why?

A: To avoid getting caught.

B: And why does he want to avoid getting caught?

C: Because he *believes* that if he is caught, he will somehow be punished.

A: What does "being punished" mean?

B: Being punished means having pain inflicted on you or being deprived of something.

C: *That* belief can be taken apart, too. First of all, there has to be a person who might be punished. Second, there has to be something he might be punished for. Third, there has to be a person or group with the *power* to punish. Fourth, there has to be an action by the punisher that produces pain or deprives the person being punished of something.

A: You know, hiding is a way of doing things we learn as children, to avoid being punished by our parents.

B: Yes, and we still use it as adults.

C: Now you're getting down to cases. Why do we use it as adults?

A: That's a pretty big "why"!

B: Well, what are some of the things we hide from each other?

C: H'mmm...I pass.

A: Coward!

B: You know, I have an impression we're all hiding! (All laugh)

C: O.K., let's list some of the things we hide. First, we hide a lot of our inner thoughts and feelings such as daydreams, feelings of pride, sexual feelings and desires—Wow!

A: We hide things we feel would be regarded as wrong.

B: Such as what? Let's get specific here, now.

C: Such as wanting to have sex with another man's wife.

A: You kinda have sex on the mind, don't you?

B: That's a topic in itself.

C: Or doing little things against the law.

A: Or not doing a job right.

B: What do you mean by "right"?

A: The way it's supposed to be done.

B: Who decides this?

A: The boss, I guess.

B: And what gives him the right to decide?

C: Food for thought, there. Another thing we hide is something we feel others might disapprove of.

A: Why?

B: I guess we just naturally want to be approved of by others.

C: But there are times when we don't give a damn!

A: There's another side to this. If I disapprove of something you have done, I usually hide that disapproval.

B: Even my best friends won't tell me! Now, why would you do that?

C: Let's watch the track here.

A: One reason would be—and this doesn't apply just to you, but to every friend—that I don't want to hurt your feelings.

B: Why not?

A: You might get mad and hurt my feelings!

C: You know, every time we hold a belief because we want to, there appear to be two motives: a socially acceptable motive and an underlying real motive.

A: You're right! And we automatically give the socially acceptable motive and hide the real one.

B: Maybe we should all try to give real motives here.

C: I'm agreeable. You know, you feel a lot better when you do this, at least I do.

A: I do, too. There's something about being honest that makes you feel good.

B: I'm agreeable, too. But I'm not always sure I can give the real motive right away.

C: You're being honest. But we can always give the socially acceptable motive first, and then look for the real one and if it's there, give that.

This example of Group Tracking could be continued indefinitely. But enough has been given to communicate the basic idea. As the reader probably has sensed, this particular topic has a wealth of material potentially available.

The reader may also have noticed that the group started out gingerly, rather abstractly and intellectually, but that as they warmed up to each other they began to get more specific. And as this warming up continued—as group rapport rose—a sense of communion, of fellowship, of brotherhood, began to emerge. Together with this was a growing awareness of deeper motives and the "good" feeling that comes from being honest.

These trends can be carried to their logical conclusions. When they are, we get a picture of a group of individuals who communicate fully and frankly, even about their most private feelings, who are ready to question anything, and who have achieved a relationship of high rapport that is rarely achieved even by the closest friends.

And that is Group Tracking at its finest and best.

It's a daring thought, isn't it—being *totally* honest with your fellow men and women. And of course it could not be done in everyday society without strong repercussions. But it can be done under special conditions such as the Human Study Groups of synergetics.

And when it is done, those who do it make the same wonderful discovery that all of us active in synergetics have experienced: *That underneath the masks we wear, each of us hides the same yearning to be a brother/sister to his fellow man/woman, to love and to be loved.*

And oh, what a heart-warming discovery this is!

CHAPTER 14

THE SYNERGETIC WORKSHOP

After a Human Study Group has been functioning for awhile, it often wishes to get together with other Human Study Groups for a period of intensive work in synergetics. This is accomplished in a Synergetic Workshop.*

A Synergetic Workshop may be of any size from about eight to fifty people. More than about fifty makes the Workshop unwieldy; fewer than eight tends to reduce its effectiveness. The optimum size is about ten to thirty.

The duration of a Workshop is limited by the time and money its members can invest. A long weekend is the minimum required to achieve good results; a week is better. Desirably it should be held in a place where food and board is inexpensive, depending on the financial resources of the participants.

The primary purpose of a Workshop is intensive work in synergetics. This involves a busy schedule from early morning to late at night, with time off only for eating and sleeping. Even at meals, workshoppers talk shop.

To be successful, most workshoppers should already know a great deal about synergetics—they should not only have read this book but also have done some Group Tracking. A small fraction, say ten percent, of workshoppers may be inexperienced in Group Tracking, but they should at least read the book before the workshop starts.

A small committee needs to take responsibility for organizing the Workshop. This is best done by an established Human Study Group, most or all of whom plan to attend. The first step is communication— getting in touch with other Human Study Groups and individuals who may be interested in a Workshop, to find out who may come, when, and for how long. A time and place for the Workshop is then determined, based on responses to this communication, and announced well in advance.

The Workshop itself should be flexibly organized. Some organization is necessary in order to use the available time and

---

*The first Synergetic Workshop was held in Columbus, Ohio in July, 1956.

resources effectively. Provision for change, however, should be incorporated into the program, and flexibility should be emphasized. Above all, workshoppers should be given the opportunity to participate in decision-making and to suggest changes.

This can be done at an organizational meeting, held at a preannounced time and place. At this meeting, workshoppers may begin by each introducing himself briefly, stating his purpose in coming to the Workshop. Then a member of the organizing committee may present a proposed schedule to the group. After discussions and questions, with opportunity for changes, the group then decides on the schedule. The meeting may then conclude by holding a Group Tracking demonstration, either by the sponsoring Human Study Group or a group especially organized for the purpose.

The Workshop itself usually consists of coaching sessions alternating with Group Tracking sessions. To accomplish this, it is necessary to assign workshoppers to groups and to co-coaching teams. For the co-coaching teams, it is desirable to ask for preferences and to emphasize that anyone can change coaches if he wants to, after the first coaching session. It should be pointed out that people who are close friends or who like each other do not necessarily work well together in co-coaching teams, and that no one should feel hurt or annoyed if his partner wants a change. In some Workshops, teams are "scrambled," i.e., entirely new co-coaching teams are formed midway through the Workshop.

The Coach's Guide should be emphasized; a good idea is to run off copies for each workshopper.

Each tracker is in charge of his own case and may choose whatever work method he wishes. For beginner coaches, it is usually a good idea to start with Operation Traverse (see Appendix). This requires no skill from the coach—all he does is ask the queries as listed. As he does so, and listens to responses, and observes and evaluates, he will often find new queries to ask flashing into his awareness. As long as the tracker agrees, he should ask them. Coaching is a lot easier than a beginner may realize, and it's fun! The reason it is easy is that each person has a lifetime of experience to draw upon, which gives him a knowledge of human beings that is greater than he could learn in books.

Indeed, coaches soon get so good at it that they get cocky and think they can do more than they really can. The tracker is in charge of his case; he is also responsible for the progress he makes. If he isn't ready to do the necessary work or to be completely self-honest, the cleverest coach in the world can do little or nothing.

Workshops are easy to organize and to conduct, and they are *very* rewarding. People often—perhaps usually—come to a Synergetic Workshop feeling a bit anxious and wondering if it isn't going to be a huge waste of time and money. You can feel this apprehension at the organizational meeting and the first Group Tracking session. By the end of the first day, a remarkable change has already occurred— people are smiling and joking and feeling great. The "Workshop spirit" has emerged.

Everybody works at a Workshop. Everybody gives. And everybody benefits. It's a very rewarding experience.

CHAPTER 15

SYNERGIC TEAMS

A synergic team is defined as a small group of human beings having these characteristics:

1. Every person in the group is a stable.*
2. The group as a whole operates in the synergic mode.

The fact that each member of the team is synergic makes it easier for the team as a whole to operate in the synergic mode and vice versa.

Just as it is difficult to describe accurately the synergic mode in individuals, so is it virtually impossible to describe the communications, interactions, and actions of a synergic team.

So much goes on so fast! So deep is the love and trust and esteem of each for the others! So beautiful is the way each instantly aligns his actions to fit the goals and interests of the others and of the team as a whole! So gaspworthy is the speed and skill with which each anticipates and facilitates the acts of his teammates!

A stable is more able than he was before; when the synergic mode turns on, a higher level of function comes into being, a holistic level that did not exist previously, and which cannot be fully appreciated until it has actually been experienced. Inevitably and naturally he accomplishes more than he would otherwise, and enjoys it more. Still, he is in no sense a superman (indeed, he is more than ever conscious of his humanity and full of compassion for his fellow humans). His abilities are limited by his natural talents, knowledge, and training. He is vulnerable—he can be hurt. He can be injured or become ill; he grows old and dies.

Quite apart from these natural limitations, a stable still has a heavy dysergy load to bear. There are, first of all, the impedances of others with whom he is closely associated. At the least they slow him down; occasionally they may cause him to react, transiently. Second, there are twangles in the groups to which he belongs; aside from the General Esteem Loss points these produce, each twangler may try to involve him in the twangle as an ally. Third, there are sociodynes—

---

*This is not absolutely necessary, provided those who are not yet stable can operate in the synergic mode.

Identic and Reactive patterns-of-expected-conduct—that are embedded in the social matrix, forcing him to do certain things and prohibiting others. Many of these are not only dysergic in themselves; they actually penalize synergic action—giving in a spirit of love is taken advantage of; consideration of another is rewarded by a kick in the face; efforts to right social injustice are met by furious oppression by the Establishment.

All this dysergy makes it more difficult for a stable to act synergically at all times; it is like trying to run at high speed at the bottom of the sea. Despite this, a stable still accomplishes a great deal more and is far happier using the synergic mode.

How different the situation is when stables get together and form a synergic team! Here, a synergic act is not penalized but rewarded. But more than this is the wonderful way that group synergy catalyzes the flow of synergy in each person's mind and vice versa. Each stable becomes part of a new entity, a Homo Gestalt. Yet he does not lose his individuality in the process; on the contrary, his uniqueness is actually enhanced!

The potentialities open to a synergic team are tremendous. Yet, paradoxically, no ordinary group or individual has any reason to fear a synergic team, because the team operates in the synergic mode. This means it will always give due consideration to the honest needs and interests of others.

There are various ways in which synergic teams may form. As more and more stables emerge and get to know one another, they may decide to form a synergic team, for example. Another is for a Human Study Group to evolve into a synergic team; indeed, this was one of the goals suggested for a Human Study Group in chapter 12. Let us assume the latter route, and consider what might be done to facilitate the process.

First and foremost is that each member of the group make a dedicated effort to become a stable as quickly as he can. Stabilization Procedure, in the Appendix, is a systematic way to accomplish this; it takes an honest, thorough, sustained work effort but the results are worth it.

Second, as soon as feasible, the group should transform itself from a true group into a cleared group. This means the elimination of all twangles, overt and covert.

Third, the members of the group should review chapter 5, expecially the synergic team functions described there. For convenience, these are listed here:

Affinity Make
Empathy Make
Semantic Telepathy
Synapse
Franktalk
Totaltalk

One or more group sessions might be devoted to these for the purpose of learning them better and developing skill in their use. *The more thay are used, the more closely the function of the group will approximate that of a synergic team.*

Totaltalk, in particular, is an emergent that is evoked by the habitual use of Affinity Makes, Empathy Makes, Semantic Telepathy, Franktalk and Synapse, combined with the practice by each syngeneer of orienting to the Broad Band. The flow of communication is so rapid and effective that it fills all who participate with joy and excitement. *Joining minds in Totaltalk* is a process that just does not happen in ordinary groups. It is truly a mind-transforming experience; one who enjoys it is never the same again.

One of the results of joining minds in Totaltalk is the emergence of an Empathic Communion. This is a relationship of such high rapport, such a remarkable degree of mutual understanding, esteem, and love, that each is virtually a part of all the others. When the Empathic Communion emerges, a synergic team can act as a coordinated unit with a minimum of spoken words, which is simply incredible. Each knows in advance what the others will do and adjusts his own actions to help, automatically and effortlessly. It is a relationship of loving/caring/sharing that not only fulfills, but transcends the innermost dreams in the hearts of men and women.

What synergic teams will accomplish cannot even be imagined at this time. But we can indicate two possibilities. One emerges from a recognition of the tremendous need, the vital importance of developing more stables and more synergic teams as quickly as possible. A team of five stables, for example, might decide to monitor the synergic evolution of, perhaps, a hundred individuals. If something like this happens, the slow task of developing synergetic stables will be greatly accelerated.

Beyond this, there is the prospect of a *group* of synergic teams organizing itself. Such a group is called a *hypergroup.* The tremendous potential of synergy applies at all levels of organization and complexity of a system. A hypergroup is as far beyond a synergic team as a synergic team is beyond an ordinary group.

One way a hypergroup might organize itself is according to the principle of pyramiding echelons.* Each team chooses one stable to represent it at the hypergroup level of organization. The stables so chosen then get together and form a new synergic team, which coordinates the actions of the basic teams. This is participatory democracy at its finest and best.

The principle of pyramiding echelons can, of course, be repeated at still higher levels if desired. While we cannot foresee in concrete detail the outcome of this process, we do gain a vision of the possibility for a synergic evolution of humanity. It is a beautiful vision. And it provides a measure of hope for the future. For such an evolution, if it occurs rapidly enough, will lead to the development of a synergic power comparable in magnitude to the enormous power of dysergy that holds humankind in thrall.

Let us hope that it happens in time.

---

*Thanks, Reynolds Moody.

PART FOUR

*Social
Synergetics*

"The Machine," they exclaimed, "feeds us and clothes us and houses us; through it we speak to one another, through it we see one another, in it we have our being. The Machine is the friend of ideas and the enemy of superstition; the Machine is omnipotent, eternal; blessed is the Machine." And before long this allocution was printed on the first page of the Book, and in subsequent editions the ritual swelled into a complicated system of praise and prayer. The word "Religion" was sedulously avoided, and in theory the Machine was still the creation and implement of man. But in practice all, save a few retrogrades, worshipped it as divine. Nor was it worshipped in unity. One believer would be chiefly impressed by the blue optic plates, through which he saw other believers; another by the mending apparatus; another by the lifts; another by the Book. And each would pray to this or to that, and ask it to intercede for him with the Machine as a whole. Persecution—that also was present and all who did not accept the minimum known as "Undenominational Mechanism" lived in danger of Homelessness, which means death, as we know.

—E.M.Forster, "The Machine Stops" (1)

CHAPTER 16

THE MACHINE
AND DYSERGY PRIME

Every human being is enmeshed in the vast and intricate world-wide system that I call the Machine. The Machine is a network consisting of a large number of social, political, military, economic and cultural components. I call it the Machine because its mode of operation is so very much like the smaller machines man has produced in such abundance. The essential characteristic of a machine is *repetition*—it repeats a sequence of movements over and over again, always the same, though differing perhaps in speed or force produced. The pistons of an automobile engine move up and down, up and down, endlessly. The wheels of a locomotive turn around and around. A printing press goes through the same complex sequence of movements, over and over again. To mechanize a process is to organize it to repeat itself as exactly as possible.

So it is with the World Machine. It imposes on each human being a variety of patterns that he or she is expected to follow repeatedly. Get up, make breakfast, wash the dishes, make the beds, go shopping, come home, fix dinner, wash the dishes, watch TV, go to bed, etc....- patterns we repeat again and again, patterns we follow exactly lest we be punished. The Machine rules us all. We are only cogs spinning around and around.

Each of us was drawn into the Machine the minute he or she was born. None has any choice in the matter. True, there are moments in our lives when we do have a choice between alternatives; but the alternatives that are available are almost always those defined by the Machine, they are seldom our own. Even here, more often than not, the Machine exerts subtle pressures upon us, manipulating us with carrot and stick to choose the alternative preferred by the Machine. Having the illusion of choice, we then feel more committed to the path we have "chosen."

How did this Frankenstein's monster arise? And why do human beings so meekly submit to it?*

What follows is admittedly an oversimplification and involves some speculations that are difficult to prove or disprove. A far better analysis is provided in the works of Lewis Mumford (2) and other writers. But it does provide a simple, clear picture that is, I believe, not in contradiction to what is known; a picture that clearly shows how we are chained to the Machine. Some might call it a fable, the Fable of the Machine. But it is a fable that rings true.

It began a long time ago. Some say about 8000 B.C.

Some unknown genius, probably a woman, discovered or invented agriculture and, at about the same time, animal husbandry.

The ways of human kind were revolutionized, never again to be the same. For the first time, humans had a reasonably assured food supply. No longer were they forced to live a nomadic existence,

---

*Ordinary machines like automobiles, washing machines, or electric typewriters operate at the command and under the control of human beings. Why shouldn't the Machine also serve man, instead of the other way around?

hunting, fishing, and picking and eating wild fruits and vegetables. No longer did they have to move from place to place, always going to where the food was. No longer were they entirely dependent on the vicissitudes of weather, climate and animal competitors for the available food supply.

Humans could grow their own food in abundance. And "civilization" became possible.

Many humans became farmers. But at this point another element entered the picture. Farmers were able to produce more than they themselves actually needed. In other words, a surplus was possible. This meant that not all humans had to be farmers.

Moreover, a farmer was *necessarily* tied to the land. He could move about, of course; but he always had to return to the farm where his crops and animals were, and he had to spend most of his time there. This left him vulnerable.

And since not all humans had to become farmers, some did not. Those who did not were those who naturally preferred to hunt and to fish and those who were good at it. Those were the ones with the weapons.

And so the hunter became the warrior. He killed the farmer and stole his food or forced the farmer to give him food under threat of death.

And in due course, some warriors were better users of weapons than others, and, true to their kind, they forced the lesser warriors to do their bidding. And they organized the warriors into armed groups. Such groups were far superior to individual warriors.

And so the warrior-kings emerged. And they took control of the land. They let the farmers continue to farm, provided they obeyed the warrior-kings and gave each king and his generals and his warriors food.

And the food the warrior-king took was called "taxes." And the orders of the warrior-king were called "laws." And the farmers and other subjects obeyed the laws "or else." And they paid the taxes "or else."

And as time went on, the warrior-king became more clever, not only in using weapons and in organizing armies, but in persuading people to obey him. He learned the warriors could be *conditioned* to obey (this is now called military training). After a period of this conditioning, the warrior learned to obey orders instantly, that to question them was to die. And he was given a false image to live up to—the image of the Hero, the Fearless and Brave. And, of course, the warrior-king was the Great One, the Champion of His People, the Superhero. Sometimes he claimed to have Supernatural Powers—to be a god.

And the same methods of conditioning were applied to farmers. The warrior-king, who had actually seized power by force, claimed that he had it by Right. And he gave the farmer another image to live up to—that of the Loyal Subject, the Obedient Citizen who faithfully obeyed the Law and paid the Taxes. And he was taught to revere the warrior-king (this is now called patriotism), who was his Protector (against other warrior-kings), and who settled his disputes with other farmers Fairly and with Justice.

And since women were physically weaker than men, but necessary for the pleasure of men and the production of new warriors, it was only fitting that women become the property of men. And so it was decreed that women belonged to their fathers or their brothers or their husbands. And the men agreed that this was Fair and Just.

Now the warrior-king was the Greatest One among his people. And this made him very happy. And since happiness equals Greatness, he naturally reasoned that he would be even happier if he were still Greater. But to do this he had to overcome another warrior-king to prove he was Number One.

Thus war was invented.

To prove his Greatness, the warrior-king invented The Enemy—another warrior-king and his soldiers and subjects.

The Enemy is always less than human, capable of murder, torture, rape, and the most unspeakable crimes. Moreover, he is out to get *you*—the Good Guy, the Hero. So you have to kill him first, before he kills you.

(Of course, the other warrior-king is doing the same thing with his young Heroes. To them, *you* are The Enemy—less than human, capable of murder, torture, rape, and the most unspeakable crimes.)

And so Warrior-King A—the Good, the Wise, the Just, the Protector of His People—orders his young Heroes to a place where they must kill the Enemy—the young Heroes of Warrior-King B. And lo! It is true! The Enemy *is* trying to kill the young Heroes and therefore he must be Evil, less than human, and everything the Warrior-King has said.

And so the young Heroes come to hate The Enemy. And the parents and sisters and younger brothers and grandparents of the young Heroes learn to hate The Enemy even more. And the warrior-king is satisfied, because hasn't it been proved that what he said was Right and True? And doesn't this prove that he is Wise and indeed the Protector of his people? And doesn't this prove that The Enemy must be murdered or tortured if he is captured? And doesn't this prove that The Enemy's women, who are less than human, deserve to be raped by the young Heroes...etc. etc. etc.?

And so the young Heroes become what they hate.

But there is one thing worse than The Enemy without and that is The Enemy within, the Traitor.

For it happens that, ever so often, a person sees through all this nonsense and refuses to take part in it. But this means that he has Disobeyed the King—the Wise, the Good, the Great. He who is not With Us, the Good Guys, is Against Us. He is one of Them and therefore less than human. Worse, he has deceived us into trusting him, pretending falsely to be one of us Good Folk.

Punish the Traitor! Kill him! Kill! Kill! Kill!

And so it went. The warrior-king was able to expand his territory and kill so many of the other warrior-king's soldiers that The Enemy was subdued. Thus, the warrior-king brought Peace to his people.

And since all the people wanted peace, and never really wanted to hate and kill in the first place, they were grateful to the warrior-king. For didn't he Protect them from The Enemy? And hadn't he ended the War and brought Peace?

Thus it began 10,000 years ago. This is the way kings and laws and taxes and governments were formed. And the people were conditioned to believe that all these things were Good and Right. And so they taught their children, who in turn taught their children, and so on down the generations until it was our turn. And we accepted it, too.

Let us pause, for a moment, and analyze the Fable of the Machine, as told thus far.

First, just as it was useful to make a distinction between individual and group consciousness, so it is helpful to distinguish a third form of consciousness, the consciousness of a person when he thinks of himself as a member of a large social unit such as a nation. I call this *social consciousness*. When an individual's social consciousness turns on, he or she may become aware of, and subject to, contents that go back many years in history. These social contents are not part of his individual consciousness and may be a source of dysergy for him.

Second, an important feature of the individual's social consciousness is *his perception of the social consensus.* This perception may be mistaken, but as long as he holds it, his actions and communications and interactions are influenced by it. These actions, communications, and interactions with others in turn help to determine the (actual) social consensus. A cyclical process is thus established, which may continue indefinitely.

Third, the Mode Ladder may be applied to social consciousness. This means, among other things, that certain processes of social consciousness may be Identic or Reactive in mode. I call such processes, and the patterns that govern them, *sociodynes.*

Sociodynes, clearly, are a major source of dysergy. This dysergy is of two main kinds: social dysergy and individual dysergy. Social dysergy takes the form of wars, crime, political and economic strife, racism, poverty, etc. The individual dysergy produced by sociodynes results from the *patterns-of-expected-behavior* they impose on the individual.

A sociodyne is much more complex than a protodyne or a chronic reaction, so much so that it requires a major study to characterize one in detail. For our purposes, it is sufficient to identify a sociodyne and perhaps to indicate a few salient features.

In the Fable of the Machine, we can identify at least three: the war sociodyne, the state sociodyne, and the male chauvinist sociodyne.

From the standpoint of social synergetics, clearly a long-term objective is to eliminate all sociodynes from the social matrix. Desirable though this may be, it is equally clear that this is a formidable task; one that would probably require many decades to accomplish. For sociodynes have been accumulating since the dawn of history and probably long before then.

From the standpoint of the individual, the patterns-of-expected-behavior associated with sociodynes are of interest. Two responses to these patterns are worth noting: the individual may accept the pattern or he may react to it.

When he accepts the pattern, his response is primarily Identic in mode; when he reacts to the pattern, his response is primarily Reactive. In either case, the response is dysergic. *And in both cases, he can clear the dysergy by eliminating his own identifications and reactions.*

This process is called *neutralizing* the sociodyne. Please note that neutralization does not clear the sociodyne from the social matrix. Also, it does not eliminate the problems and difficulties the individual may encounter in dealing with the sociodyne in its social and group aspects. But the individual does have the power to neutralize sociodynes, and when he does so, he feels better and can achieve more.

Let us now return to the Fable of the Machine.

As we have seen, the Machine was born when agriculture was invented. And since farmers were able to produce a surplus, not all people had to be farmers. Some became warriors, as we have seen; and the warrior-kings emerged and established the State. Others became carpenters, weavers, smiths, masons, etc. producing goods and services that people needed or wanted. At first, these were traded by simple barter. Later, money was invented as a medium of exchange and a measure of value.

By and by another group of people began to emerge. Workers and farmers were very busy working and farming. They had little time to take their goods to people who needed them or to find where these people were. So some people saw in this an opportunity. They bought the goods cheaply from the farmers and workers who produced them, and they stored them or took them to people who needed the goods and sold them for high prices. These were the traders.

Now some traders realized they were performing a service and charged for that service only what they needed to buy their fair share of the goods and services that society as a whole produced. But others didn't worry about this. They had no qualms about cheating or deceiving the people they bought from or sold to. Buying cheap and selling dear enabled them to keep the difference, which they called "profit." And some traders became very rich this way.

One day, a rich trader got the idea of bringing all the workers together in one place called a manufactory. This made it easier for the trader, who could organize the workers so they produced more. And since the trader was very rich, he could pay the workers for the goods they produced, only now he called this payment wages. And since the workers now had to get wages in order to buy food and other things they needed, they pretty much had to do what the trader wanted. In this way, the trader became a boss.

Some workers remained free, of course, at least for awhile, producing things themselves and selling them. But free workers could not compete in the long run. Because things were organized in the manufactory, goods could be produced a lot more efficiently and sold at a lower price than the goods of the free worker, even though the trader-boss still made a large profit.

Meanwhile, science was invented. A scientist is a funny person who is smart in some ways and dumb in others. Scientists began to find out more and more about nature; and one day they discovered how to make machines run by natural energy.

The trader-bosses were delighted about this. Being very rich, they paid the scientists to design machines for their manufactories; and they paid workers to build the machines. With machines, workers could produce a lot more than they could before. And so the manufactories became factories. And the trader-bosses became even richer.

Some of the trader-bosses were generous and kind and tried to help the workers to make things easier for them. But to do this cost money and reduced their profits. So the trader-bosses who were most ruthless, buying the workers' labor cheap and selling their products dear, made the most profits and soon drove the generous trader-bosses out of business.

Sometimes a factory became so efficient that it produced more than people wanted or could afford to buy. When this happened, the trader-boss would lay off workers in order to keep his profits.

The trader-boss didn't mind this, because it made sure that the workers he kept on would work harder and not get any radical ideas.

And the trader-bosses saw the value of machines and the scientists who were smart enough to design them. So they hired scientists (who were then called engineers) to design more and better machines. And most of the scientist-engineers didn't mind this, because they loved to do research and design machines, and the trader-bosses paid them well. They became workers, bought and paid for like any other thing by the trader-bosses. But most of them didn't realize this.

Trader-bosses were constantly trying to figure out new ways to buy cheap and sell dear. Some of them noticed that money could also be treated as a commodity to be bought cheap and sold dear. They also saw that farmers and workers needed a place to keep their money, temporarily, until they spent it.

So these trader-bosses became money-bosses. They stored the

money of the workers and farmers in a safe place called a bank. But they realized that on any given day farmers and workers would draw out only a small fraction of their money; the rest of the time the money-bosses could use it as they pleased. So they loaned it out, mostly to other trader-bosses, and charged interest for it. And in this way they became very rich.

Now the trader-bosses and the money-bosses and the modern-day warrior-kings (who are called politicians) are not evil men. They are just doing what their roles tell them to do. If they don't perform according to their roles, the Machine causes them to lose, and they become workers or farmers or unemployed.

The Fable of the Machine could be continued indefinitely; but enough has been told to provide a basis for understanding how this Frankenstein's monster arose. Let us now consider the second question posed at the start of this chapter: Why do human beings so meekly submit to it?*

A simple answer would be: Because we have no choice. But this may immediately be modified to: Because we *believe* we have no choice. And this leads to another question: Why do we accept this belief?

Whenever the Machine permits us the illusion of choice, it always defines the alternatives in terms favorable to it. Catch 22. In the case of submission to the Machine, the choice offered is submit or perish. You need food, which only the Machine can provide. And, for almost everyone, this is true. But there are other alternatives, which become clear once we refuse to limit ourselves only to those stated by the Machine.

But there is more to the problem than this. Let us probe more deeply.

The human mind is so organized that when a pattern occurs that "works," the mind tends to use that pattern again in a similar situation. This is the Identic mode in operation. There is no discrimination or awareness.

The warrior-king established the original pattern of the state with laws and taxes. This same pattern prevails today, despite the fact that,

---

*There are, of course, individuals who love the Machine. For those, the question has no meaning. We leave them to their worship.

to many thinking persons, the nation-state has become an anachronism, and despite the fact that the original warrior-kings used brute force to seize power and to keep that power. *Whatever was, is right;* and it will be repeated, over and over again, till the end of time.

The endless repetition of the Machine is nothing more than the projection, upon the screen of social consciousness, of the Identic mode of function.

This insight provides the basis for understanding why the Machine has such power over us. It is because *we unconsciously give it that power.* It is as if the Machine were under the control of a pseudo-mind, operating entirely in the Identic and Reactive modes. Like the Freudian Id, which imposes its protodynes to control the perceptions, thoughts, feelings, and actions of the individual consciousness, this pseudo-mind imposes its sociodynes upon our social consciousness.

It will be useful to give this pseudo-mind a name. I call it Dysergy Prime because it is a primary source of dysergy upon the planet earth.

To understand how Dysergy Prime controls us, let us once again consider the distinction between individual and social consciousness. When you are alone, doing something you enjoy and do well, your consciousness is your own; it isn't shared with anyone else. The contents of your consciousness are *your* sights, sounds, ideas, etc. Your will, too, is your own—you do as you choose, selecting from alternatives you yourself formulate, not those of the Machine.

But when you become a member of a group, your consciousness subtly changes. The contents of your consciousness are no longer purely your own, but are selectively focused upon those contents that you perceive to be occupying the consensus-attention of the group. Moreover, you tend to perceive those contents the way the group-consensus perceives them, which may differ from the way you would view them through individual consciousness.

The same thing applies to your will. You are limited to alternatives that you perceive to be acceptable to the group, indeed, that are expected by the group. You may, on occasion, choose; but you do so as a member of the group, not as an individual.

Now, in addition to *membership* in a group, an individual also has one or more *roles* in the group and a certain *status* in the group. If your role happens to be one of leadership, you may imagine that you have power to shape things as you choose and, to a limited extent, you do. But power attracts power seekers like garbage attracts flies. You soon find yourself surrounded by flies clamoring for attention or cleverly massaging your ego so they can wield some of the power in your name. And in your role, you are expected to manage every crisis that comes along. Your consciousness again is not your own, but is governed by your perception of the contents occupying group attention, modified by your perception of what-you-are-expected-to-do.

If your status is low in the group hierarchy, your image of yourself is profoundly affected. Instead of seeing yourself as you actually are— a unique individual with unique potentials and unique needs—you view yourself in terms of your status, your perception of how-you-are-regarded-in-the-group. You feel inferior, inadequate; and because this is unpleasant, you may try to compensate by covering it up, by taking it out on someone else whose status is even lower than yours or by finding something outside yourself to glorify. As for your will, your lowly status insures that you end up doing the scut work that nobody else wants to do.

Generalize this to include all the groups to which we belong, and to society as a whole, and we see that we become absorbed in a social consciousness that to a considerable degree governs what we perceive, the way we think, the way we regard ourselves, and the actions we take. Each of us assumes a False Identity that is a kind of average of his various memberships, roles, and statuses, while the real "I" is submerged, confused, and impotent.

But this is not the whole story. From social consciousness that occupies our minds there emerges a new entity whose existence and reality are determined by collective agreements. This entity is Dysergy Prime.

Dysergy Prime may be defined as the set of sociodynes that have accumulated since the beginning of humankind, controlling us through patterns-of-expected-conduct associated with our memberships, roles, and statuses in various groups and other social entities. The Machine serves Dysergy Prime, not individual human beings. The Machine itself is only a machine, and could be made to serve us if Dysergy Prime were destroyed.

How can Dysergy Prime be destroyed?

In principle, the solution is easy. Dysergy Prime exists because we unconsciously believe in it. If all of us, collectively, became fully and knowledgeably conscious of Dysergy Prime and of the ways it controls us, and collectively decided not to believe in it any more, Dysergy Prime would disappear.

Unfortunately, this has to be a collective decision. One person, discovering the truth, can stop believing in Dysergy Prime; but his social consciousness discloses that everyone else still believes; and their collective belief so influences their perceptions, thoughts, and actions that his individual consciousness is overwhelmed. He must continue to deal with the reality of this collective belief.

In the movie "Forbidden Planet," a human scientist, Dr. Morbius, spent many years studying the remains of an advanced civilization whose citizens had mysteriously disappeared. The resources of the planet had been harnessed by a technology far in advance of that of man, and this technology continued to function automatically. Ultimately, Morbius discovered that the vanished race had been destroyed by an Id creature produced by their collective unconscious, a creature that drew upon the inexhaustible energies of that technology.

Let us hope that a similar fate does not await humankind.

Whatever you can do, or dream you can, begin it.
Boldness has genius, power, and magic in it.

——Goethe

CHAPTER 17

SYNERGIC POWER

The Machine has grown to such an enormous size and complexity, and its influence over the individual is so overwhelming and stultifying, that it is easy to understand why some people despair for the future of humankind. The difficulties confronting us are indeed great, with potential catastrophes looming on many sides. Nevertheless, there are good grounds for hope.

First and foremost is the fact that we now have a basic understanding of the problem. The Machine, itself, is not the problem; under appropriate conditions, the Machine may be transformed into a machine serving man instead of controlling him. The problem is Dysergy Prime, the unconscious collective projection of the Identic Mode upon the social consciousness of humankind—the "idiot mind" in control of the Machine. Knowing the existence and basic nature of Dysergy Prime, we can study its characteristics and predict its behavior in given circumstances. And as our knowledge of Dysergy Prime grows, we will develop methods to counteract its power, to contain it, and ultimately to reduce it to impotence as a factor in human affairs.

It would be asinine to say that this will be done easily or soon, or that we now have the tools to do it. And at any moment, one of the looming catastrophes may destroy us all. But paralysis and despair in the face of danger will not eliminate the danger; no one knows when such a catastrophe will occur, and we now have a road to travel that will enable us ultimately to prevent these catastrophes—all of which will be caused, if at all, by Dysergy Prime! The existence of these looming catastrophes gives urgency to our tasks and zest to the adventure of accomplishing them.

A second reason for hope is that we now realize that no single economic class, political ideology, race, nation, or other social entity is to blame for the human predicament. All are equally under the control of the Machine and Dysergy Prime. It is easy to find villains and scapegoats to account for our many problems, but in doing so we are ourselves unwittingly operated by sociodynes, without our awareness, knowledge, or consent. The capitalists are not to blame, nor the communists, no more than the blacks, the poor, the Arabs, or the Jews. As we learn to become aware of sociodynes and to neutralize them, we can well be guided by the words of the poet:

> *He drew a circle that shut me out—Heretic, rebel, a thing to flout.*
> *But Love and I had the wit to win: We drew a circle that took him in.*

——Edwin Markham

A third reason for hope is the recent development of some powerful new ideas and tools in social synergetics, largely by Craig and Craig (1) and Hampden-Turner (2). Let us now consider these developments.

Two Types of Power

Traditionally, power has been viewed as "the ability of an individual or group to carry out its wishes or policies, and to control, manipulate, or influence the behavior of others, whether they wish to cooperate or not." (3) Craig and Craig characterize this form of power as *directive* power. It consists of five elements:

1. A power wielder—the initiator.
2. An object or target of power—the responder.
3. A goad or punishment—the "stick" with which the power wielder coerces the responder.
4. An incentive—the "carrot" with which the power wielder rewards the responder.
5. The behavior path—the activity of the responder desired by the power wielder.

This is the familiar carrot-and-stick concept of power, and there is no doubt that it is the prevailing form of power in use today. The amount of power a power wielder has is determined by his capacity to reward or punish the responder. Sometimes he emphasizes punishment; at other times, reward. He can be cruel or gentle in the way he exercises power, and the rewards and punishments used may take many forms. But it seems clear that all directive power can be characterized in terms of these five elements.

The Craigs point out that another type of power exists, which they call *synergic power*. It is the kind of power exercised by Jesus, Gandhi, and Martin Luther King. The Craigs define synergic power as "the capacity of an individual or group to increase the satisfactions of all participants by intentionally generating increased energy and creativity, all of which is used to co-create a more rewarding present and future." (1)

Synergic power differs from directive power in that coercion is not used. The carrot and the stick are laid aside. "Synergic power differs radically from directive power in the concern it expresses for other people and the roles it affords them. Any application of synergic power accords with the will, the judgment, and the interests of other human beings, and it is fully effective only when no energy and creativity is wasted in domination and resistance to domination." (1)

The Craigs introduce what they call a "technique-free" model of power to encompass both directive and synergic power. Here we shall simply use this model to analyze the elements of synergic power. These elements include:

1. An initiator who is at the same time a responder.
2. A responder who is at the same time an initiator.
3. An environment consisting of three parts.
   a. A shared environment, common to both parties.
   b. A unique environment of the initiator.
   c. A unique environment of the responder.
4. Two-way communication between the initiator and the responder.
5. Behavior in the form of response/work by both parties.
6. An outcome that produces satisfactions for both parties.

It is easy to see how this model can be used as a basis for generating synergic power. The initiator starts things off; his immediate goal is to increase his satisfactions by influencing the behavior of another party (person or group). In doing so, however, he is ready to relate to the needs and aspirations of the other party and to respond to them in turn. He is also aware that, in addition to their shared environment, the responder has a unique environment of his own, with a context of potentials and concerns unknown to the initiator.

Instead of using carrot and stick to coerce behavior, the initiator uses communication—two-way communication. The Craigs propose a four-element model by which the initiator may synergically influence the responder's motivational state. These are:

1. The responder's conceptual map of the way things are.
2. The responder's conception of things as they might be if changes were made—alternative futures.
3. The responder's conception of possible ways in which a desirable alternative future might be brought about.
4. The responder's evaluation of the efforts and risks involved in achieving such an alternative future.

It is clear that communication must be two-way if the initiator is to understand and to influence the responder's motivational state. He has to start from where the other is at. And while the initiator has his own individual or group goals or satisfactions that he wants to fulfill, the degree of synergic power he achieves will be determined, not by his capacity to reward or punish the responder, *but by his ability to match the desired behavior path of the responder to the responder's motivational state.*

This model of a synergic power transaction is diagrammed below. A clear understanding of this diagram and the Craigs' four-element model of the responder's motivational state will enable the syngeneer to generate and to exercise synergic power to a remarkable degree, even if, from the standpoint of directive power, he is impotent.

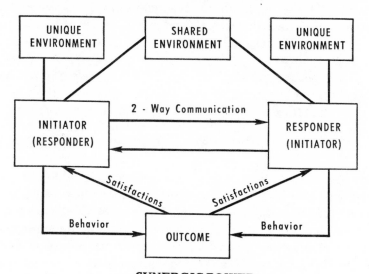

SYNERGIC POWER
(Adapted from Craig and Craig) (1)

There is one vitally important caveat: *synergic power and directive power are mutually exclusive.* A person who uses directive power on one occasion will find it very difficult to use synergic power on another. This does not mean that it cannot be done (see below); but he will find that the responder is suspicious, perhaps even hostile. The past exercise of directive power against him leads him to expect that it will be so used again. And even if he assures himself that it will not, he may be predisposed to take advantage of the situation to "get even."

This does not mean that a syngeneer should never use directive power. The structure of human society imposes on nearly everyone, at one time or another, roles he is constrained to perform. A higher directive power is used to coerce him into using directive power on those "below" him in the power hierarchy, whether he likes it or not. A parent is responsible to society for the behavior of his child; if synergic power does not work, he will himself be punished if he fails to use directive power. There are also emergency situations, when a person in a position to use directive power feels obliged to use it to prevent injury or harm to others; there simply isn't time to use synergic power.

Another situation in which the use of directive power is sometimes warranted is when a responder uses directive power against a power wielder. Many power wielders become enamored of their power; they need it to bolster their weak egos—it is their security blanket. A power addict is relatively immune to synergic power; his needs and satisfactions all involve the exercise and enhancement of his directive power. As long as the responder uses directive power for defense and not revenge, such use is justified.

It is almost always possible, however, for a syngeneer who has been compelled on one occasion to use directive power to remedy the situation afterward. Even if he cannot eliminate the dysergy entirely, he can usually reduce it. Even if he cannot reduce it, the effort should be made. Later, the fact that he has made the effort may count in his favor.

Many persons have used synergic power intuitively, without having analyzed what they are doing. The Craigs, however, have provided us with a clear and well-thought-out model that enables us to use it with loving precision. Synergic power is the ideal antidote for impotence; no amount of directive power can prevent its use or cancel its effectiveness. But more than this: synergic power is the basic technique that will enable us to create waves of synergy that can not only touch everyone we meet, but can spread out with synergic effects in all directions. Each of us who uses synergic power becomes a synergy source; and as more and more of us generate synergy, and our efforts coalesce, we will bypass the centers and nodes of directive power and achieve a synergic world. We cannot predict when this will happen or what form it will take; but we can know that if we persist long enough, synergy must prevail. For synergy touches and fulfills the deepest aspirations of humankind.

Power Struggles and the All-Win Method

The concepts of directive and synergic power provide us with tools for understanding many social phenomena. It is at once apparent that modern "civilization"—the society of every nation of the planet—is based on an edifice of directive power. This is as true of the "free world" as it is of the "socialist camp." It is as true of the nations of the "Third World" as it is of the "developed" world. These directive power structures take different forms, and appealing ideologies are often used to deceive, confuse, and divide the responders to keep us impotent and duly responsive. But power structures and hierarchies are everywhere, and they dominate the affairs of humankind.

The wielders of power are not to be blamed for this state of affairs. They are themselves the unwitting puppets of Dysergy Prime, playing the roles demanded of them. Let us widen our circle of love to take them in, even as we see through their efforts to deceive, confuse, and manipulate us.

Directive power being what it is, it is inevitable that power wielders come into conflict with one another from time to time. Whenever this happens, a power struggle ensues. Each power wielder then uses his directive power to compel his responders to carry out the struggles. Such power struggles are found everywhere, but especially in the worlds of politics, diplomacy, and economic enterprise. Sometimes they take violent forms; when this happens on a large scale, we have war.

Not all power wielders are power addicts, however; and not all conflicts are between power wielders. The Craigs have developed a promising method for dealing with such conflicts. It is called the All-Win program.*

---

*A somewhat similar approach is used by the SYNCONs sponsored by the Committee for the Future. (4)

The basic idea behind the All-Win program is to transform adversary relationships, which usually dominate intergroup conflicts, into a synergic relationship where everybody wins. The key to the success of an All-Win program is the use of synergic power to establish five basic conditions that will permit cooperation and participation to flourish. According to the Craigs, "These are: (1) an atmosphere of openness and growing trust, (2) a shift of focus from abstract principles to what is actually happening to real people, (3) replacement of adversary relationships with cooperative problem-solving relationships, (4) a free flow of reciprocal information between responsive representatives and the informed people they represent, (5) a climate of cooperative creativity nurtured by coaches and group members trained to use synergic communication, and by a commitment to revise agreements should they at a later point prove unsatisfactory." (1)

The All-Win program involves three main elements:

1. A core conference consisting of open meetings between representatives of the various parties to the conflict. Proceedings are communicated via closed circuit TV or other methods.
2. A network of citizens' groups spread throughout the community, each group meeting regularly to discuss core conference meetings and feed back their responses.
3. All-Win Coaches, one for each team of delegates to the core conference.

Each coach should himself have some knowledge of synergetics or related fields; behavioral science professionals would be natural for this role. Each coach would help his team "identify the human concerns hidden beneath abstract arguments" (1), and guide it in the core conference.

It is strange, at first, for people accustomed to adversary relationships to participate in processes designed to create conditions in which everybody wins and nobody loses. But as each party becomes aware of the hidden human concerns that underlie the other parties' positions, a degree of mutual understanding begins to emerge. As this grows, each party is better able to trust the others, and a realization emerges that energy spent in conflict can also be used to create cooperative processes, with far greater benefit to all concerned.

The Craigs emphasize the importance of having all agreements reached be open-ended. This means "a joint commitment to re-examine the issues and create new solutions should the present agreement later prove unsatisfactory to any side. Thus, one serious difficulty commonly experienced in trying to resolve conflicts is greatly reduced: the difficulty of overcoming politicians' natural inclination to keep all options open." (1)

The All-Win method, in my opinion, probably will not work if one or more of the parties to the conflict is a power addict. And it will be on very shaky ground if it is sponsored by a party having directive power over all the others, unless there is a prior understanding that this directive power will not be used. With these caveats, the All-Win program looks very promising.

An Application of Synergic Power:
Ten Principles of Human Development

The potential applications of synergic power are legion. Once the basic idea is clearly understood, the ingenuity of the human mind may be counted on to derive techniques appropriate to each concrete situation. As more and more people master and apply this idea and other ideas and tools of synergetics, we may expect synergic power techniques to emerge in delightful profusion.

One especially useful application of synergic power has been advanced by Charles Hampden-Turner. (2) This takes the form of ten principles of human development. Hampden-Turner develops these principles in the context of a strategy for poor Americans; but the principles themselves are quite general and may be used by any oppressed group, class, etc.

It is impossible to do justice here to Hampden-Turner's work; for this, the interested reader should consult his writings. (5) Here, I shall simply outline his ten principles and indicate in a general way how they may be used.

1. The principle of free existence. By this is meant the right of a member of any group, class, race, religion, etc. to originate and to choose. Actually, any human being has the basic power to do this; but *as a member of an oppressed group* he often finds this power restricted and its exercise punished. Indeed, he may find it so taken for granted that he *lacks* the right to free existence—due, perhaps, to some presumed defect of character—that he accepts the belief that he does not have it or, that in order to get it, he must conform to some standard set by others.

To facilitate the recognition and assertion of this right, Hampden-Turner proposes the creation and development of organizations that generate economic surplus value and originate social and political programs. Through such organizations, the individual may counteract the stultifying effect of the institutions that oppress him and discover that he does have the right to exist as an originating and choosing being—and the power to exercise that right.

2. The quality of perception principle. "There is great *definitional power* as well as creativity in the right to label reality and have that label stick." (2) When this definitional power is used to define events in a way that is derogatory to an oppressed group, members of that group are compelled to respond in terms of the way those events are defined. Such phrases as "ghetto violence" or "the Negro problem," when applied to particular events, emphasize aspects of those events that are unfavorable to blacks. The same events might equally well be defined as "reactions to intolerable conditions" or "the white problem."

By the quality of perception principle is meant "the capacity to see what *is* and the imagination and creativity to see potentials and ideals within the status quo and within oneself that could transform the situation." (2) To facilitate this, Hampden-Turner proposes that members of oppressed groups develop "paradigms of their own, that is, consistent and coherent ways of perceiving contradictions and articulating plans to resolve them." Institutions and organizations are needed to enable such groups to set their own goals in terms of needs as *they* perceive them, not as some benevolent Great White Father defines them.

3. The strength of identity principle. A member of an oppressed group, regarded by others with hostility, suspicion, scorn, etc.—or with pity and condescension—naturally tends to doubt himself at times or to become overly preoccupied with defense at others. This tendency is reinforced by the status he is accorded by social institutions and by the roles he is called upon to play. His unique individuality is submerged in a sea of social invalidations and given little or no opportunity to develop. The wonder is that it does develop, even under such adverse circumstances.

To compensate for this, Hampden-Turner proposes the creation and development of organizations and institutions that provide positive, self-fulfilling roles to its members. Such roles, by enabling the individual to use his abilities in ways he himself perceives to be useful and beneficial to his community, reinforce his self-esteem and enable him to compensate for the icy winds of derogation that assail him as a member of an oppressed group. "They permit poor people to fill positions of skill, variety, mobility, and importance, roles which stand in marked contrast to those of 'welfare recipient,' 'dependent child,' 'problem family,' 'hard-core jobless,' and 'high-risk, target area resident.' " (2)

As important as the provision of high-quality role structures is the presence of role *mobility*. The institutions and organizations should provide ample opportunity for the individual to move from one specific role to another that is within his capabilities but at the same time challenges his further development.

4. The principle of competence. Knowingly or unconsciously, most people experience a feeling of impotence vis à vis the Machine. This feeling is greatly heightened for a member of an oppressed group, who sees himself stymied at every turn and all too often retreats into drugs, crime, or insanity.

To compensate for this, Hampden-Turner proposes the development of structures that promote a *synthesis* of competences. Once more the concept of synergy applies: where one individual or group has competence in one area, and another in a different area, the two together can achieve more than they can separately. In union, there is strength. This is an old idea but it is a powerful one.

5. The principle of authentic and intense commitment. "The test of social action is whether, in fact, one's true EXISTENCE, PERCEPTION, IDENTITY, and sense of COMPETENCE has been COMMITTED through the act...Only through such COMMITMENT, with the emotion aroused through real involvement, can we learn to reconcile head and heart, mind with body, and discover the fusion of abstract values and concrete experiences." (2) Or, to put it more succinctly, there must be think-act synergy.

To facilitate this, Hampden-Turner advocates the development of "emotionally significant opportunity structures" (2) that provide avenues for meaningful participation by members of oppressed groups. He emphasizes that this participation must be more than a nominal "rubber-stamp" or "response to questionnaire" type of participation; that the individual must clearly have a share in decision-making procedures. The opportunity must be real if the commitment is to be wholehearted.

6. The principle of suspension and risk. By this is meant that the individual opens his mind and heart to new experience and perspectives whenever he commits himself to actions in which the outcome is uncertain. In other words, he temporarily suspends the assumptions he makes about the world he encounters. Necessarily, this suspension involves some risk. "I discover that I am valuable and meaningful to the other, only by risking the discovery that I am worthless and meaningless." (2)

If suspension and risk is too great, however, the individual becomes so exposed and vulnerable that the opportunity for development is overwhelmed by his or her anxiety. This is especially the case in encounters between black poor people and government welfare agencies. Hampden-Turner points out that "almost every government agency, if it can, will devise a 'no-fail' strategy so that it is justified in either or all events...the poor individual is 'chronically needful' if he falls within the Agency's categories, or a 'welfare chiseler' and 'drain upon the public purse' if he does not." (2)

To compensate for this, Hampden-Turner proposes establishment of an organizational structure that "controls degrees of uncertainty and vulnerability in daily life." (2) Through collective action, risks can be reduced to the point at which the potential reward is seen as worth the risk taken. The rich long ago discovered how to do this via the corporation, an organizational structure that limited their liability. There is no valid reason why comparable organizations could not accomplish the same for the poor.

7. The principle of bridging the distance. When two groups encounter each other, each perceives the other as different from itself. The degree of difference perceived is called the distance between the two groups. According to this principle, development of both groups occurs when this distance is effectively bridged.

To facilitate this, Hampden-Turner proposes the establishment of a structure that "builds formal alliances and balances the ratio of differentiation to integration." (2) When different groups freely ally themselves in order to achieve a common goal, the distance between those groups is effectively bridged. The bridging is even more effective when the differences between the two groups are utilized in a division of labor that optimizes the contribution of each to an integrated action.

8. The principle of self-confirmation and self-transcendence. By self-confirmation is meant the understanding and appreciation by others for one's actions and achievements. Self-confirmation provides a social boost to self-esteem. By self-transcendence is meant the confirmed knowledge that one's actions and achievements have contributed significantly to the group or community. Clearly, human development is promoted by appropriate amounts of self-confirmation and self-transcendence.

Hampden-Turner proposes to facilitate this by establishing a structure that "confers social benefits, fuses social and material rewards into a symbolic system, and becomes a repository of meanings." (2) He emphasizes that such a structure must be quite different from the consumer training provided by the mass media. "There issue from TV sets and magazines invitations to indulge, spend, lick, snuggle, bask, masticate, and devour." (2) Such titillations are inducements to behavior that the poor cannot afford and that is contrary to the hard work and struggle that are essential to their development.

9. The principle of dialectic leading to synergy. The dialectical process is a continuous one, involving encounters between groups whose aims are in apparent conflict, but which are reconcilable by means of interactions leading to a new, synergic perspective within which the aims of both are achievable. Hampden-Turner believes this may be facilitated by means of a structure that provides an "arena for conflict resolution between laterally structured, interdependent groups requiring a multiple synthesis of ends." (2) He emphasizes the importance of balance, social justice, and equality between the parties involved if a synergic resolution is to be achieved.

10. The principle of feedback ordered into complexity. Hampden-Turner does not consider these principles of human development to be a linear array of ideas added together to form a composite whole. Rather, he views them as intercalated parts of a cyclical process. As each principle is applied, the process of development proceeds up to the higher synergy of the ninth principle, which is fed back "into mental matrices of developing complexity." (2) This is the way a person grows—not just as an isolated individual, but as a member of an evolving community—unless that growth is somehow interrupted.

To promote this feedback, he suggests the creation of "a system of authoritative and legitimate expectations based on neighborhood sovereignty" with "a marshalling of community resources." This structure should bring about "a record of shared experiences and expectations, and an historical continuity which imparts a sense of moving forward through time, and passing on to one's children's children an accumulating body of knowledge." (2)

Hampden-Turner uses these ten principles of human development to develop a strategy for poor Americans that emphasizes their self-determined evolution through organizations and institutions of their own choosing. Key components of this strategy are the Community Development Corporation (CDC) and Social Marketing. The CDCs are local organizations having economic, social, educational, and political functions. Social marketing is a new institution—"the offering of goods for sale which are labeled and advertised, not on the spurious grounds of miracle ingredients and acquisitive advantages to the consumer, but on the basis of the humanity, the community, and the social purposes behind the products...The strategy of social marketing is to create a supplier-consumer alliance between the poor organized into CDCs on the one hand and, on the other hand, relatively affluent and college-educated young, organized into groups, which give allegiance to peace, civil liberties, ecology, social justice, women's rights,...etc." (2)

It's an intriguing strategy. For further details, Hampden-Turner's book should be consulted. Here I want to emphasize that these principles of human development are perfectly general and may be applied in a variety of ways to a variety of situations and groups.

Two applications in particular stand out. One is to the analysis of inadequacies and problems in the development of any group, class, race, nation, etc.—or even to humankind as a whole. It seems probable that such inadequacies may be understood and remedies for them derived in terms of one or more of these principles.

The second is to the creation of a truly synergic society. The dead weight of dysergy is so great in every nation on the planet that it may well be that only a new beginning, a fresh start, will enable humankind to establish the social conditions necessary for actualizing the human potential. At the least, such a move merits serious consideration. This will be the task of the next chapter.

There is a revolution coming. It will not be like revolutions of the past. It will originate with the individual and with culture, and it will change the political structure only as its final act. It will not require violence to succeed, and it cannot be successfully resisted by violence. It is now spreading with amazing rapidity, and already our laws, institutions and social structure are changing in consequence. It promises a higher reason, a more human community, and a new and liberated individual. Its ultimate creation will be a new and enduring wholeness and beauty—a renewed relationship of man to himself, to other men, to society, to nature, and to the land. This is the revolution of the new generation.

——Charles A. Reich (1)

CHAPTER 18

TOWARD A SYNERGIC SOCIETY

How can we develop a society with high social synergy, described so well by Ruth Benedict as one whose "institutions insure mutual advantage from their undertakings"?

No complete and final answer can be given to this question. The answer, when it comes, will be the work of many minds and hands, acting in synergic alignment, each using the special talents, knowledge, skills, facilities, and resources available to him or her. What I shall try to do in this chapter is to outline a few of the lines of action that are now clearly open to us. Hopefully, as we move along these lines, new vistas and opportunities will emerge. This is one of the delightful attributes of synergic action. As we view these vistas and act on these opportunities, the concrete form of the new synergic society will unfold like a flower. Of one thing we can be sure: there will be many surprises.

There are some things each of us can do entirely alone, with only this book and our own knowledge and experience needed to guide us. Other actions will require joint efforts. Some actions can be undertaken at once. Others will require facilities and resources that are not now available to us. As we grow and develop, and create or gain access to these facilities, we will do these things. We will act not according to some preset plan, but in the manner of a living organism: starting small and growing quietly—and exponentially.

Project Mutation

The first and foremost task for each of us is to turn on the synergic mode in our own minds and to stabilize there.

Some readers may already have achieved this state without calling it that. If you have, welcome, comrade! You know how wonderful this state is and how vital it is for as many as possible to achieve it as quickly as possible. We need your help and will align our actions synergically with yours. Please contact us soon.

If you have not, and do not consider it desirable or necessary to do so at this time, we understand and respect your position. We can still work together toward developing a synergic society if you are so inclined. The same applies to those who would like to become "stables" but are not convinced they can do so. We believe it is less difficult than you may think, but perhaps later you may change your mind.

The remainder of this section is addressed primarily to those who would like to become stables and want to know how.

The Appendix of this book contains an organized procedure that we believe enables a person of at least average intelligence and self-honesty to stabilize in the synergic mode. It is called Stabilization Procedure. There are several ways in which it might be used.

1. *Solo Track.* An individual working alone can stabilize in about fifty to 100 sessions of about one hour each. This is the simplest and most direct method. It requires a working knowledge of synergetics (such as that obtained by reading this book twice and studying relevant parts), and an honest, thorough, sustained work effort.

2. *Teamwork Track.* Two people, working together, can stabilize together—each in turn coaching the other using Stabilization Procedure. This takes longer, but the probability of success is higher. This is especially so if good rapport develops so that each reinforces the work effort of the other.

3. *Group Track.* A group of about four to ten people using Group Tracking to supplement solo work and/or teamwork, can not only stabilize together, but also form a synergic team in the

process. This is probably the best and surest method. It does, however, require that someone invest time, energy, and thought into the organization of a group. But the results are worth the effort.

4. *Synergy Centers.* The foregoing methods are relatively simple and require no formal organization; and costs are nominal. Clearly, however, a good deal more could be accomplished with greater efficiency by an organized effort on a larger scale. This can be achieved by the establishment of Synergy Centers.

Professionals in the fields of medicine, religion, mental health, education, the behavioral and social sciences, and the human potential movement are admirably qualified to organize and operate Synergy Centers. There are many ways in which this might be done, and what follows is not intended as a rigid prescription, but as a guide.

Let us assume that a clinical psychologist wishes to organize a Synergy Center without giving up his practice. How might he proceed?

First, it would be desirable to incorporate the Center as a nonprofit educational institution whose purpose would be to teach people the art and science of synergetics. This procedure will vary considerably from state to state. It is desirable, however, to provide legal protection for the Center and for its organizer or organizers.

A second step is to apply for income tax exemption as a nonprofit organization from the Internal Revenue Service. This usually takes time, but it is a highly desirable step. This enables the Center to receive contributions that are tax-exempt for the donor.

Meanwhile, the organizer should seek access to facilities adequate for conducting Stabilization Courses (described below). These facilities do not have to be elaborate, but should include a number of rooms, or subdivisions of a larger room, within which individual sessions could be held. A large old house or a church might be utilized for this purpose.

There are (at least) two kinds of Stabilization Course that are feasible.

The first, a Regular Stabilization Course, is designed for the individual who wants to stabilize without giving up his usual activities. It is estimated that such a course (based on Teamwork with some Group Tracking) would require 150 to 200 hours. At five hours per week, this would take thirty to forty weeks. It is highly desirable to be flexible about this, because some people will be able to stabilize faster than others. It is also desirable, however, to set a maximum time for the course, after which it will be regarded as concluded whether the person has stabilized or not.

The second, an Intensive Stabilization Course, is designed for the individual who wants to stabilize as quickly as possible, and who is ready, able, and willing to devote full time to it for a limited period. The total time required will be about the same—150 to 200 hours—but at fifty hours per week this could be completed in about three to four weeks.

Applicants for these courses should be carefully screened to insure that individuals with severe neurosis or psychosomatic disorders are handled separately. Such individuals need psychotherapy and are unlikely to be ready to handle the sometimes tough and searching queries of Stabilization Procedure. On the other hand, Stabilization Procedure may be very useful for patients who are in the terminal phases of psychotherapy; such persons often have achieved greater self-honesty than "normals" and may zip through Stabilization Procedure much faster than normals.

A word about costs. Tuition for these courses must, of course, be sufficient to cover these costs, and it is certainly appropriate for those who teach and administer the courses to receive a fair compensation. There is, however, a long tradition in the synergetic movement that no one be charged a fee for coaching he receives. The distinction between coaching and psychotherapy has an *economic* side as well as a *technical* side. I sincerely hope that this tradition will continue to be observed.

As a precaution against unethical individuals who might seek to commercialize Stabilization Procedure, the Synergetic Society will establish and maintain a list of approved centers that agree to incorporate as nonprofit organizations, to use the Stabilization Course and other synergetic techniques approved by the Society, and to operate in an ethical manner.

Synergy Centers will not be restricted to Stabilization Courses and are expected to evolve, especially as more and more stables are enabled and as more and more synergic teams emerge.

The Jupiter Project

What will stables and synergic teams do?

First, they will continue their own synergic evolution. There exist advanced modes of being beyond that of stable; indeed, we have yet to discover limits to the degree of synergy that can be achieved. Techniques for enabling these advanced modes are under development and will be made available to stables as they emerge. As for synergic teams, they will take charge of their own evolution, made possible by the Empathic Communion that emerges in such groups.

Second, they will participate in activities in support of individuals and groups that are seeking to prevent the "looming catastrophes" that threaten us all such as nuclear holocaust, mass starvation, explosive overpopulation, ecocatastrophe, overwhelming pollution, etc. These activities are referred to for convenience as the Jupiter Project.

Third, they will engage in activities and processes designed to develop synergic power for the ultimate purpose of eliminating Dysergy Prime. This is the Ossian Project, discussed in the next section.

The Jupiter Project is not in any sense a formal, organized project. Nor is it necessary to predict in detail what the stables and synergic teams will do. Some activity is already underway, although necessarily on a small scale. Based on experience to date, a few characteristics of the Jupiter Project are beginning to emerge.

1. Stables and synergic teams do not advertise themselves as such. Rather, they proceed quietly, using synergic power. Indeed, the actions and mode of being of a stable can only be fully perceived and understood by another stable (or by someone who is temporarily operating in the synergic mode). A person unfamiliar with synergetics will be completely unaware that a stable is at work. He will simply observe a quiet, likeable person who gets things done.

2. Stables do not, in general, seek or operate from positions of directive power. Occasionally, they may have directive power thrust upon them, in which case they assume the role easily; but even here they characteristically proceed by synergic power. Only in an emergency will they use directive power, and they always clear up any dysergy produced by such action as soon as feasible.

3. Stables act characteristically in one of two roles: as catalyzers and as facilitators. Often the ingredients for a synergic action are present in a situation, but need a spark to ignite them. When a stable perceives this, he provides the spark. This may require some initiative on his part, but it is done in such a way that once ignition occurs, the stable quietly withdraws into the background. This is catalytic action. It has a high gain-to-effort ratio.

In other situations, a problem or obstacle occurs that interrupts or slows down action. Sometimes the problem is unsolvable with the individuals and resources available, but more often all that is required is for someone to perceive the problem and its solution and to take action to put the solution into effect. Sometimes it is simply a matter of formulating a clear line of action that will solve the problem and communicating it to the parties involved. On other occasions a twangle may be imminent. The stable distracts the attention of the twanglers from one another temporarily, maneuvers to achieve a separation, and then monitors each to help blow off steam. While doing this he analyzes each party's BAMs and VIPs for the purpose of finding one or more areas of potential agreement, areas that can be characterized in concrete terms. He then brings the parties into communication again, and using an area of agreement as a bridge and the Synergy-Empathy-Communication Triangle as a tool, evokes improved rapport between them. Effort invested in twangle-prevention saves a great deal of time, energy, and thought in the long run, that might otherwise be absorbed in dysergy.

In this and other ways the stable acts as a facilitator, expediting movement toward group goals with relatively little effort. It is a role that a stable is uniquely qualified to perform.

4. Finally, one of the most effective things a stable can do is to evoke traverse. Estimating the mode of a group, he is able to communicate and interact with the group in such a way as to enable it to climb to the next step of the Mode Ladder, and then the next step and so on until at least the Multiordinal Mode is reached. He accomplishes this not by "lecturing" the group or by advocating a particular viewpoint or by giving advice. Rather, he listens and facilitates emotional discharge; and when appropriate, he asks queries *requiring thought to answer*. The queries asked should not be too searching or require total objective self-honesty—such queries tend to evoke reactions and diminish rapport. Rather, the stable suits the query to the mode. His objective is to turn on the rational mind of the responder. Gradually, a step at a time—always permitting the responder to come up with his or her own solution, never imposing one preferred by the stable—he enables the responder to climb the Mode Ladder.

Only when the Multiordinal Mode is reached may the stable assume a more active role. If he does, it will be primarily to introduce a *synergic perspective* that integrates relevant viewpoints and evokes think-feel synergy in the group.

These, then, are some of the ways in which stables are operating in the Jupiter Project. Of course, the effectiveness of two or more stables can be greatly amplified if they operate as a synergic team and even more as a hypergroup. However, it would be a mistake to assume that stables, synergic teams, and hypergroups will be able to prevent the "looming catastrophes" by their actions in the Jupiter Project alone. All they can do is to reduce the probability that any of these catastrophes will actually occur. In the long run, more will be achieved by a given effort through the Ossian Project. This, however, is by no means obvious.

The Ossian Project

The ultimate objective of the Ossian Project is the elimination of Dysergy Prime.

Dysergy Prime is a very abstruse target. It cannot be seen or heard or touched. It is not located in any physical place. It cannot be directly observed with the aid of even the most sensitive of physical instruments; no microscope or telescope is powerful enough to make it visible. It exists only in the minds of men and women.

Even this existence is profoundly different from other mental contents. Dysergy Prime is buried in the minds of each of us and is activated when our social consciousness turns on. Its survival is made possible by the unconscious projection of the Identic and Reactive modes upon the screen of social consciousness. When a group is aware of this process, *it is possible for them to clear Dysergy Prime from their zone of interaction,* provided they are sufficiently isolated from other individuals or groups. When this is done, the group is said to be in an O-Zone.*

The establishment of an O-Zone involves far more than the neutralization of sociodynes, and it is possible only to a group functioning in the synergic mode, for practical purposes, a synergic team. In an O-Zone, each member is aware of the mechanism by which Dysergy Prime is produced—the unconscious projection of the Identic and Reactive modes upon the screen of social consciousness. He knowingly "cancels" this projection, basically by traversing up the Mode Ladder to synergy whenever it operates.

This means that he is aware of the sociodynes that tend to operate

---

*It is possible for a stable to establish an O-Zone by himself. However, the real advantages of an O-Zone are best obtained in larger zones.

upon him: the patterns-of-expected-conduct associated with his membership, statuses, and roles. He does not conform to them (Identic Mode). Nor does he react against them (Reactive Mode). *He moves to synergy with respect to them.*

The really vital process involved in creating an O-Zone, however, is the *social* process involved—everyone on the team does it knowingly, and at the same time *helps* everyone else on the team with this task. The end result is the complete elimination of sociodynes from the social consciousness of the team!

When this is done, there emerges an almost explosive expansion of awareness, freedom, and potential. Awareness is free of the highly attenuating impact of sociodynes upon the minds of the individuals in the group. But even more remarkable is the vast expansion of freedom. It is like being at the bottom of a well and then emerging from the well onto the earth's surface. The freedom of the individuals in an O-Zone far exceeds the freedom previously available in any community or society on this planet. Only in an O-Zone can a person really appreciate the degree and extent to which human beings are chained by sociodynes.

As for the increased potential, this, too, is far beyond that available in ordinary society—so far beyond that it appears to be limitless. Ultimately, there must be limits, but they are not yet discernible.

The creation of O-Zones is the key element in any effective strategy for eliminating Dysergy Prime from the social consciousness of humankind. Beyond this, our present vision is too restricted to enable us to see what can and will be done; and indeed it would be unwise to try. Better to concentrate on creating and expanding O-Zones (and joining them together in a Synergy Web). Then, within their O-Zones, stables and synergic teams will be so much more aware and so much freer to develop their greatly augmented potential, that they will be far more able to plan and carry out that strategy.

It would be most unwise, however, for a synergic team to form an O-Zone without at the same time setting up an Anti-dysergy Shield to protect the team and its members. This means restricting input communication and interaction with the O-Zone to prevent an inflow of dysergy greater than the team can handle. This must be done, however, as synergically as possible. Far more important, however, are the self-restrictions that team members use when *outside* the O-Zone. "When in Rome, do as the Romans do" is an injunction that is extremely important here. The heady freedom of an O-Zone is so great that it is the most natural thing in the world for a stable to continue exercising this freedom when he is outside the O-Zone. But this is almost certain to produce dysergy. It is vital to wear the chains that humanity wears when outside an O-Zone.

End of the Beginning

From the perspective of Project Mutation and the Ossian Project, it seems clear that these projects are far more important than the Jupiter Project and will actually create the conditions necessary for the Jupiter Project to succeed. Clearly, it is desirable to enable as many stables as possible as quickly as possible, in order to provide a basis for synergic teams to emerge. Clearly, O-Zones should be created as soon as possible, and a Synergy Web connecting these O-Zones established, so that an effective strategy for eliminating Dysergy Prime can be designed and implemented. For Dysergy Prime is the real scourge of humankind. Were Dysergy Prime to be eliminated, the looming catastrophes could be readily dissolved.

But this is easier said than done, because most people do not have a synergic perspective and cannot readily achieve it without a major reduction in their dysergy loads. Only a small fraction of people will want to undertake Stabilization Procedure when first they hear of it. Were we to focus our efforts entirely on Project Mutation, progress would be slow. A purely rational line of action is rarely a socially acceptable one. If it were, we would long ago have had a democratic world government.

Enough time, energy, and thought should be invested in Project Mutation to insure that those who want to undertake Stabilization Procedure are given the help they need. Efforts to persuade people to try it are at best a waste of time and may even be counterproductive. The wiser course is to do what needs to be done to make it known and available, and to let interest develop naturally at its own pace.

When I first began to work in the field of synergetics, I was driven by a sense of urgency. It was clear that a nuclear arms race would occur, and that the time would soon come when humankind would have the capability of destroying civilization and killing hundreds of millions of people in a nuclear holocaust. That time has indeed come, and the nuclear arms race continues. To this threat has been added many other imminent catastrophes. A realistic appraisal of the prospects of the human race must lead to the conclusion that the odds are against us. Indeed, they are growing worse.

Yet, curiously, I no longer feel this driving sense of urgency. I know the threats are there, and that they are growing; and I am also aware of the danger of complacency. But I feel that synergetics enables us to do something about these problems. More importantly, I feel that synergetics is only a part of a much larger movement involving many individuals and groups in all the lands of the earth, a movement that is quietly but steadily producing a new consciousness. It doesn't matter what it is called, and the fact that many different ideas and methods are being used to evoke this new consciousness is a source of great strength. I know that synergetics is helping, and that is what counts. And as I feel this new consciousness emerge—not in the headlines or in the hierarchies of directive power, but quietly in the minds and hearts of people everywhere—I am *certain* that we shall succeed.

STABILIZATION PROCEDURE

The synergic mode is an exhilarating state of expanded awareness, enhanced rationality, and think-feel synergy that is available to the human mind. It is characterized by a high degree of synergy among mental functions. Awareness expands to encompass the broad band, with high synergy among all tracks and tempos. It also expands to include multiple perspectives from which situations are viewed and multiple goals for which actions are undertaken, again with high synergy among perspectives and goals. Rationality is enhanced through self-programmed thinking, an Information Source orientation, and total, objective self-honesty. Think-feel synergy emerges from the habitual use of Directives and Thrillbeam on synergic potentials. As a result of these and other synergies, a new, holistic level of function emerges that is greater than the sum of its parts. This is the synergic mode.

It is relatively easy to turn on the synergic mode in a synergetic session, using techniques previously described in this book, provided the individual is, temporarily at least, relatively free of dysergy. Once the session is over, however, and the individual again is confronted with his dysergy load, mode tends to drop. In order to *stay* synergic most of the time, it is necessary for the syngeneer to reduce his dysergy load to manageable proportions.

The purpose of Stabilization Procedure is to enable the syngeneer to do just this. In other words, it is designed to enable him, or her, to *stabilize* in the synergic mode—to stay synergic most of the time. Such a person is called a synergetic stable.

Please note that it is not claimed that a stable is synergic *all* of the time. In the first place, it is not feasible in a world as dysergic as ours for any person to eliminate his dysergy load entirely. There may occur moments when even a stable cannot manage his dysergy load and his mode drops. However, he quickly traverses to synergy again. In the second place, new dysergy may be encountered at any time, which, if sufficiently great, may produce a mode drop. Again, however, a stable rapidly climbs the Mode Ladder to synergy.

Stabilization Procedure consists of five parts:

Operation Traverse
Protodyne Reduction Procedure
Operation Milquetoast
Creative Procedure
Stabilization Telopause

Operation Traverse is designed to enable the tracker to open his case and to quickly gain insights into some basic relationships which, experience has shown, are often bogged down in dysergy. By responding to the queries in each Regular Session, the tracker turns on processes in his mind that evoke traverse up the Mode Ladder.

Intercalated with the Regular Sessions are Synergy Sessions, which are designed to enable the tracker to consolidate the gains made in Regular Sessions and to achieve greater synergic command of his mind. These Synergy Sessions are also included in the other parts of Stablization Procedure.

Protodyne Reduction Procedure is designed to enable the tracker to reduce the effect of protodynes—unconscious, Identic Mode patterns that, when stimulated, control a person's perceptions, thoughts, emotions, and actions. The intent is not to eliminate protodynes entirely, but to reduce them sufficiently to permit the tracker to work on Self-Invalidations.

Operation Milquetoast is designed to enable the tracker to locate and reduce Self-Invalidations. A Self-Invalidation is a turn-off of the creative evolution of a person in a particular area. When a Self-Invalidation (S.I.) occurs (or is reactivated), a protodyne is substituted by the mind as a primitive, emergency way of dealing with the situation. S.I.s are the primary target of Stabilization Procedure. When they are fully eliminated, protodynes are no longer necessary and they simply fade away.

Creative Procedure is designed to enable the tracker to systematically eliminate all S.I.s. It is a very powerful procedure, provided it is carried out thoroughly. As S.I.s are eliminated, protodynes fade away and the tracker is enabled to operate in the synergic mode almost all the time. The Synergy Sessions, meanwhile, have steadily enhanced his ability to function synergically.

After completing Creative Procedure, the tracker is "enabled." Stabilization is then easy, almost automatic. He or she discovers that it doesn't take an effort to be stable; it takes an effort *not* to be stable.

This is not an end, however, but a new beginning. It is like being born again. There is a great deal of re-evaluation to be done of past education and experience, and a re-formulation of basic life goals and aims from a synergic perspective. In addition, there are advanced modes of being beyond stable. Perhaps most exciting of all, there is the exhilarating prospect of contacting and getting to know other stables and in forming or joining synergic teams.

To facilitate these processes, and to help eliminate any residual dysergy that may have been overlooked, the *Stabilization Telopause* is available. To participate in this, contact the Synergetic Society. Current address is 1825 N. Lake Shore Drive, Chapel Hill, North Carolina, 27514.

Stabilization Procedure can be carried out in any of three ways:

1. Solo Track—the individual works alone. This is the most direct path, but it requires a sustained work effort.
2. Teamwork Track—two persons work together, each alternately coaching the other, following the Coach's Guide.
3. Group Track—a small group of four to ten persons organize a group for Group Tracking. They form teams for the Teamwork Track. In addition, they may seek to evolve into a Synergic Team.

How to Use Stabilization Procedure

To get the most out of Stabilization Procedure, and to become a stable faster, the following are recommended:

1. *Set up a definite Work Schedule.* This is very important. A given query sequence takes from thirty minutes to two hours; but it doesn't have to be finished in one session. An hour a day is a reasonable schedule. Some may find it more convenient to work intensively on weekends. The important thing is to have a definite time and place for work and to stick to your Work Schedule as closely as possible. Stabilization Procedure is for *workers,* not dilettantes. The harder you work, the sooner you will become a stable.

2. *Get a notebook or a tape recorder or both.* This is important for two reasons: (a) the act of describing in words your responses to the queries helps to bring material more clearly into awareness; (b) the record makes easier periodic reviews, which often bring new material into awareness and which help to clear impedances only partly cleared previously.

3. *Follow the Tracker's Guide.* The Tracker's Guide is a set of practices, policies, and principles that, experience has shown, facilitate work effort and help to solve problems that may arise. It is recommended that the tracker study this thoroughly and review it frequently (see chapter 10).

4. If you use a coach, be sure he understands and abides by the Coach's Guide (see chapter 11). The simplest way to coach is just to ask the queries and nothing more. Anyone can do this, and be sure he is abiding by the Coach's Guide. However, a perceptive coach can often think of additional queries pertinent to the material the tracker is producing. As the coach gains in experience, he is encouraged to ask these queries, remembering that the purpose of a query is to *evoke thought,* not to elicit a response that fits the preconceived ideas of the coach. Always remember that the tracker is in charge.

5. The queries of Stabilization Procedure are simple and usually easy to answer. Occasionally, however, a query will seem to be ambiguous or the tracker will "draw a blank" on his response. When an ambiguity is encountered, the tracker should interpret the query in the way that feels right for him, or even modify it to fit the context of his own case. (The coach may make suggestions to help with this.)

If the tracker "draws a blank"—for example, if the query asks him to recall a particular kind of incident and he cannot—don't worry about it. Ninety-nine percent of the queries are answerable; it would be expecting too much of any procedure for it to be absolutely perfect for everyone.

Sometimes a tracker can rephrase the query and get an answer— for example, he could *imagine* an incident of a particular type if he can't recall one.

6. Progress in Stabilization Procedure is not a straight line upward. It is rather a series of ups and downs superimposed on a straight line upward. Every time an impedance is encountered, mode tends to drop; and from this low position, things don't look so good. Conversely, from a high position, things may look better than they really are. A true perspective can be achieved by viewing the situation as a whole; when this is done, the tracker realizes that the general trend is upward.

Sometimes the tracker will become vaguely aware that he seems to be moving in a predetermined way, governed by a force outside his awareness or control. He may experience an emotion of sadness or guilt for no apparent reason; this may be followed by a blocking of thought or effort, and perhaps by a daydream; or he may feel tired or sleepy. Though it may not appear so at the time, *this is actually progress.* The tracker is temporarily caught in a protodyne. *But he has become aware of it!* True, his awareness is vague, and he may feel momentarily unable to do anything about it. But previously he was entirely unconscious of the protodyne. He has begun the climb upward from the Identic Mode. As he climbs higher, he learns that he *can* do something about the protodyne.

7. A given query sequence may be used more than once. The second time through will often result in producing material that is quite different from the material produced the first time through. Usually, it is better if the second session does not occur immediately after the first one, however.

8. Although Stabilization Procedure is a unit, each session (or component procedure like Operation Traverse) is complete in itself, and may be used by itself. This is particularly valuable in Synergetic Workshops where time usually does not permit going through Stabilization Procedure as a whole.

9. Finally, the important thing about these queries is their *cumulative effect.* Each query is part of a synergic sequence; each response sets the stage for the ones to follow. The response to a single query is usually not spectacular; but a small bit of work does get done, and a small step forward is made. A momentum builds up that often lasts for hours or even days after the session is over.

This momentum is precious! Even though, as noted above, each session or component procedure is complete in itself, *the best results from Stabilization Procedure occur when a tracker goes through it in its entirety, on a definite schedule.* A synergic macroprocess is activated that reinforces the work effort of each session and continues to promote progress between sessions.

So...keep tracking!

Operation Traverse

Work Session #1—Introduction

Q1- 1. What is your worst trait?

Q1- 2. What is your best trait?

Q1- 3. What is your second worst trait?

Q1- 4. Why do you regard this as bad?

Q1- 5. What is the opposite of this trait?

Q1- 6. Describe an incident in which this second worst trait showed itself.

Q1- 7. Describe an incident in which the opposite of this trait showed itself.

Q1- 8. Describe an incident in which this second worst trait occurred in somebody else.

Q1- 9. How did you feel when this happened?

Q1-10. Describe a way in which this second worst trait has a value for you.

Q1-11. How else might you achieve this value?

Q1-12. Look now, at your reason for regarding this trait as "bad." Why do you accept this?

Q1-13. Imagine what things would be like if you did *not* accept this ideal. Consider this for your work, for your love life, for your relations with friends.

Q1-14. Describe your body sensations now, in head, eyes, throat, hands, heart, stomach, genitals, and feet.

Q1-15. What is the *opposite* of the ideal (Q1-12 and Q1-13)?

Q1-16. Describe an occasion in which you acted in accordance with this opposite and felt justified.

Q1-17. What is your emotional tone now? (Fear, apathy, greed, exultation, anger, love, cheerfulness, etc.)

Q1-18. What would you *like to do* if this emotion were to gain natural expression?

Q1-19. Think of a situation in which you would regard such an expression as wrong.

Q1-20. What is your chief reason for wanting to change yourself?

Q1-21. Consider again the emotion of Q1-17-18-19. Think of a situation in which you would regard such an expression as justified.

Q1-22. What would you *really* like to do in such a situation, if there were "no holds barred"?

Q1-23.   What *need* prevents you from acting this way?

Q1-24.   Why do you regard this as a need?

Q1-25.   How do you *feel* about this need?

Q1-26.   What new way or ways of achieving this need can you think of?

Q1-27.   What is your chief reason for wanting to change yourself? (Take a *fresh* look.)

Q1-28.   What would you like to do now, if you could do anything you wanted to?

Q1-29.   Do you have a feeling that you *have* to change yourself? (Q1-27)

Q1-30.   Why? (or, if "No," Why not?)

End of Session

Synergy Session #1

Q-1.   Review, with the aid of your notebook, your response to Work Session #1. Search for additional material that may have only partly emerged previously.

Q-2.   Read chapter 2 of this book.

Q-3.   Sweep the Mode Ladder, looking for present time processes that may be going on at each rung.

Session Review

Work Session #2

Topic: Mother

Note: If your Mother is dead, there may be unresolved grief. It helps to permit a "grief discharge" to occur.

Introduction

When we formed our first personality, the ones who influenced us most were our parents. We had very little data when we started out and couldn't even talk or understand words, so we copied the way our parents did things. This is a natural thing to do, of course; but perhaps if we had known *then* what we do now, we would do things differently.

(The opposite of this is interesting. If we knew *now* what we knew *then*, we might also do things differently.)

For most of us, Mother was the parent who influenced us most in the early days. So, it seems reasonable to start out with Mother to study what things we do now that are "copies" of the way Mother did things.

(For trackers whose Mother died early, do this session on Mother-substitute, which may have been Father.)

Q2- 1.   What trait do you like most about Mother?

Q2- 2.   What trait do you like least about Mother?

Q2- 3.   Describe a trait that Mother wanted you to develop but you didn't.

Q2- 4.   How did you feel about this when you failed?

Q2- 5.   Describe a time when Mother punished you for doing something wrong.

Q2- 6.   How did you feel about it?

Q2- 7.   What did you try to do to regain Mother's approval?

Q2- 8.   Do you now have the trait you like most about Mother? If not, describe a trait of Mother's that you do have.

Q2- 9.   What is the opposite of this trait?

Q2-10.   Have you ever acted as if you had the opposite of this trait?

Q2-11.   Describe your body efforts now—the tendency of different parts of your body to do things—in feet, genitals, stomach, heart, hands, throat, eyes, and head.

Q2-12.   Consider now the trait you like least about Mother. Do you have this trait? If not, think of a bad trait of Mother's that you do have.

Q2-13.   How do you feel about this trait? What dominant emotion is associated with it?

Q2-14.   Think of another emotion that you sometimes associate with this trait.

Q2-15.   Think, now, of a situation in which this trait showed itself. What *specific part* of this situation acted as a stimulus for the dominant emotion?

Q2-16.   Suppose you were to encounter this same situation again. How would you deal with it now?

Q2-17.   How would you deal with it if there were no limits on what you could do now?

Q2-18.   What is the chief limit that keeps you from doing this?

Q2-19. What is the next most important limit that keeps you from doing this?

Q2-20. For whichever of these two limits you choose: Why do you accept this limit?

Q2-21. How would you feel about having sexual relations with Mother?

Q2-22. Describe a situation in which you needed Mother and she wasn't around or did not respond to your need.

Q2-23. Describe a situation in which Mother needed you, and you *did* respond to her need.

Q2-24. How do you *feel* about being controlled?

Q2-25. What do you *think* of people who are domineering and bossy?

Q2-26. Have you ever been domineering and bossy?

Q2-27. Name at least three ways in which you "controlled" Mother.

Q2-28. Describe another situation in which you needed Mother and she wasn't around or did not respond to your need.

Q2-29. Do you still have this need now?

End of Session

Work Session #3

Topic: Father

Introduction

Father stands for authority in our lives; the way we respond to authorities in general is derived from the way we responded to Father when very young.

Father also stands for strength, courage, endurance, and fighting spirit. These are qualities that are important to a human being, male or female, and in developing them we copy Father.

Later, because Dad had his failings, we noted little (and sometimes big) ways in which he did not perform according to our expectations. For each of us, this was a personal tragedy greater than we may realize, not so much because we care about Dad, but because we *needed* him, at the time, as a model for learning how to live and get along. And many of us tried to restore this lost image, and the effort so spent stopped us from growing and perhaps still does.

In this session we "turn the tables" on Dad. The subject pretends to be Father and his (or her) own Father is actually his (or her) own son. It gives a different feel to things.

Q3- 1. Take a trait you dislike in Father. How would you go about correcting this trait if he were your son?

Q3- 2. Suppose he refused to cooperate. What would you do?

Q3- 3. Suppose he goes on being ornery no matter what you do, and it makes you so made you float to the ceiling and stick there, legs pumping in angry unison. You are just furious! What would you do to this child of yours?

Q3- 4. Eliminate *all* restraints and try again!

Q3- 5. Now look at the restraints that keep you from doing this. Name them one by one, and write them down. Then, for each, answer the following queries: Q3-6 through Q3-11.

Q3- 6. Name a situation in which you tried to impose this restraint on another and it worked.

Q3- 7. Name a situation in which you tried this and it didn't work.

Q3- 8. What general belief about people or the universe did you adopt when this happened?

Q3- 9. What are you doing *now* to try to make up for having failed in this?

Q3-10. What did you need or want in the situation (Q3-7) that you didn't get?

Q3-11. In general, what other way or ways might you fulfill those needs or wants?

Repeat this exercise for one more trait, and for as many more as you wish.

End of Session

Work Session #4

Topic: Brother and Sister

You may not have had a brother or sister, but you did have playmates and can use them for this exercise.

"Right" and "wrong" derive their authority from our parent's teachings, but they gain their reality from our experiences with our brothers, sisters, and playmates. "Justice" is done when what's "right" is rewarded and what's "wrong" is punished, and, above all, it must be *equal—exactly* equal or else! Otherwise, it is unfair and we may spend the rest of our lives seething in rebellion against *something,* not quite knowing what.

For this reason, you are a judge; and everything that you did that was right but unrewarded, or wrong but unpunished, will now be set right; and the same for what your brothers and sisters did to you and to one another.

Q4- 1. Describe a situation in present-day life—either in your own life or in politics—that you regard as unfair.

Q4- 2. Who is responsible?

Q4- 3. You are a judge and can do whatever needs to be done to render justice. What would you do?

Q4- 4. Now describe an incident that happened to you that you regarded as grossly unfair.

Q4- 5. Again, you are the judge. What would you do to render justice to yourself?

Q4- 6. Now take another look at the incident and what followed. What *did* you do to render justice to yourself?

Q4- 7. What is your emotional tone *now?*

Q4- 8. Describe the efforts in your body, in order: head, eyes, nose, mouth, ears, tongue, throat, hands, heart, breathing, stomach, knees, feet, genitals, and back.

Q4- 9. Was anything ever done to you behind your back that was grossly unfair?

Q4-10. Describe the hurt you felt.

Q4-11. Now take this hurt and *feel* it in the parts of your body as follows: back, stomach, knees, elbow, hands, thumbs, eyes, teeth, genitals, nose, head, heart, breathing, back of neck.

Q4-12. Adopt the viewpoint of getting revenge for this wrong. What would you do if there were no limits on your actions to gain this revenge?

Q4-13. Intellectually, do you regard "getting even" as wrong? What is your opinion of the matter (in detail)?

Q4-14. Now adopt the viewpoint that it is necessary for you to *justify* getting even for this wrong. How would you go about doing this? (Be *real clever* on this one!)

Q4-15. Now look objectively at the parts of your body, in this order, to see whether or not a "feeling of hurt" is present (or any other feeling): wrist, elbow, tongue, eyes, genitals, stomach, heart, breathing, back, back of neck, top of head, soles of feet, throat, and breasts.

Q4-16.   Can you think of a time that you took something that wasn't rightfully yours?

Q4-17.   How did you feel about it afterwards?

Q4-18.   Can you think of a time that somebody took something that was rightfully yours?

Q4-19.   Be very reactive and vindictive now, and, in your imagination, punish the wrongdoers by blows or actions at the following parts: teeth, eyes, lips, genitals, hands, stomach, throat, shins, back.

Q4-20.   What is your emotional tone now?

Q4-21.   Now, for the time you took something that wasn't rightfully yours, imagine somebody as mean and vicious as you were is about to punish you, and describe how you would *protect yourself* against blows or actions against the following parts: back, teeth, throat, genitals, shins, stomach, eyes, mind.

Q4-22.   Define justice.

End of Session

Synergy Session #2

Q-1.   Review Work Sessions #2-4. Search for additional material that may have only partly emerged previously.

Q-2.   Read chapter 5 of this book.

Q-3.   Select one of the Work Sessions of Q-1 and do a Broad Band Sweep on contents and processes.

Session Review

Work Session #5

Topic: Sex

Sigmund Freud is generally credited with being the man who exposed the extent to which repressed sexual desires play a part in the lives of all of us.

A lot of people think he was wrong about a lot of things, including especially things he never said. This latter is interesting in its way. But anyhow, in this exercise you are going to outfreud Freud, in imagination, of course. You are going to imagine yourself as doing, or having done to you, the most outrageous sexual practices you can think of—things you would never, never, never do in real life, no matter what the circumstances...Or would you?

Q5- 1. How would you feel about having sexual relations with your mother?

Q5- 2. How would you feel about having sexual relations with your father?

Q5- 3. Now assume you just *might* have a blind spot,* and answer each of these questions again.

Q5- 4. Now let us pretend you have decided to rape a woman. You are a soldier at war and your best friend has just been killed and your turn may be next. Imagine in detail how you would go about doing it. (If you are a woman, pretend you are a man and do this anyway.)

Q5- 5. Now let us assume you are a middle-aged woman who has never had intercourse and is approaching the change of life. Imagine in detail how you would go about trapping a man. (If you are a man, pretend you are a woman and do this anyway.)

Q5- 6 Now review your rape (if a man) or your trap (if a woman)— embellish it a little if you wish—and after having done this, look at your sensations in each of these parts of your body: eyes, lips, cheeks, tongue, hands, face, fingers, buttocks, thighs, genitals, breasts, heart, stomach, and mind.

Q5- 7. Now assume you are being raped (if a woman) or trapped (if a man) as you have imagined it above, and protect yourself in each of the following parts of your body: nose, mouth, lips, tongue, genitals, stomach, back, genitals, feet, heart, and genitals.

Q5- 8. Repeat the above, except this time help yourself to *be* raped or trapped.

Q5- 9. Repeat the above and be disgusted at what you have done.

Q5-10. Repeat the above and be pleased at what you have done.

Q5-11. Repeat the above and hide the pleasure so that nobody will ever find it.

Q5-12. Now look at the places where you have hidden the pleasure and see what you have hidden there before.

---

*A "blind spot" is an area a person just doesn't look at, no matter how self-honest he is trying to be and how hard he is working.

Q5-13.   Go before a mirror and pick out the things you admire about yourself. Admire them.

Q5-14.   Now pick out the things that you regard as ugly or not admirable. Feel revolted about them.

Q5-15.   Look again at what you admire and cover up your self-admiration. Do a good job. Be *real* modest.

Q5-16.   Look at your ugly points (might as well call a spade a spade) and cover them up.

Q5-17.   Name three areas in which you have not been honest with yourself.

Q5-18.   Name three areas in which you have been *conspicuously* honest.

Q5-19.   Look again at your admirable features and admire them again. Now, how would you go about bringing them to the notice of one of the opposite sex without seeming to?

Q5-20.   Think of a situation in which you have tried to keep up appearances.

Q5-21.   Think of a situation in which your self-esteem was hurt and you tried to restore it.

Q5-22.   Look again at your ugly parts (don't forget the back) and figure out how you would go about justifying them.

Q5-23.   Think of two occasions in which you tried to live up to an ideal and failed.

Q5-24.   For each, how did you justify your failure?

Q5-25.   Think of two occasions in which you did something that was good, and you knew it but nobody paid any attention.

Q5-26.   For each, how did you go about attracting attention?

Q5-27.   Look at the mirror again, and if a male, go through all the motions of being strong and masterful, etc. If a female, go through all the motions of being irresistible and precious.

Q5-28.   Do the same, only switch sexes.

Q5-29.   What is self-honesty?

End of Session

Work Session #6

Topic: Reality

Q6- 1.   Have you ever been so angry you were beside yourself with rage?

Q6- 2.   Pretend you are a ferocious tiger and are about to kill a deer. Imagine—and illustrate with gestures—how you would go about it.

Q6- 3.   Describe a situation in which you were completely baffled, didn't know what do do, and *had* to do something.

Q6- 4.   Imagine you are inside the sun, but immune to its rays. Now imagine that you *are* the sun. Warm everyone with your beneficent rays.

Q6- 5.   Describe a situation in which you were taken advantage of.

Q6- 6.   Now imagine that you are God, all-powerful, all-wise, all-knowing.

Q6- 7.   Describe a situation in which you were completely ignored.

Q6- 8.   Now imagine that you *are* the universe, all of space and time, the planets and stars and people and animals and plants and atoms and molecules.

Q6- 9.   Describe a situation in which you were ridiculed and laughed at.

Q6-10.   Now imagine that you are in the middle of the sun again and the heat becomes unbearable and you evaporate.

Q6-11.   Now imagine yourself evaporating out to the limits of the universe until you *are* the universe again, only this time the universe is somehow closed and it closes in on you until it goes right through you and you're still there.

Q6-12.   Describe a situation in which you successfully attracted attention and admiration.

Q6-13.   Now imagine that you are the Devil, evil incarnate. What would you do to these admiring people?

Q6-14.   Describe a situation in which people thought you couldn't do something but you did.

Q6-15.   Now imagine that you are a mouse and a cat is after you. It draws closer and closer and catches you, and chews you up, mashing your bones, and swallows you, and dissolves you in its stomach, and you become nothing.

Q6-16.   Describe a situation in which you successfully used anger to get something done.

Q6-17.   Now imagine that you are inside an insane asylum, and everyone around you is crazy but you. Imagine some of the things that might happen and that you might do.

Q6-18.   Describe a situation in which you successfully used fear to escape a real danger.

Q6-19. Pretend you are a blood corpuscle, and start out in your heart, move out through the arteries to the capillaries, back through the veins to the heart. Do this again, moving from heart to brain and back. Do it again, moving from heart to foot and back.

Q6-20. Describe a situation in which you were afraid you might die.

Q6-21. Pretend you are a master hypnotist and can make anybody do anything you wish. Imagine a scene in which you hypnotize everybody and get everything you want.

Q6-22. Describe a situation in which you successfully escaped being punished.

Q6-23. Describe a situation in which you successfully got your way over all opposition.

Q6-24. How do you know you are sane?

End of Session

Synergy Session #3

Q-1. Review Work Sessions #5-6. Search for additional material that may have only partly emerged previously.

Q-2. Read chapter 7.

Q-3. Select one of the Work Sessions of Q-1 and do a Sweep of Determinants.

Session Review

Work Session #7

Topic: Grief

Everyone has, at one time or another, experienced grief. Being human, there are people and things we learn to love; and sometime or other we lose them. The capacity for love and the capacity for grief are but two sides of the same coin. You cannot have one without the other.

Practically everyone also has a certain amount of unresolved "grief work," too. Even those who have "worked on" their grief by one method or another may still have it.

Unresolved grief cripples our capacity to love freely, generously, spontaneously, and creatively. If you are unable to do this, in every situation, you may have grief work to do.

This session is different from the others; it is done without a coach. The subject goes somewhere where he can be completely alone, probably out-of-doors. He then takes the following query sequence and repeats it for his father, his mother, his brothers and sisters, his grandparents, his aunts and uncles, and everyone he has loved.

Q7-1.   What don't (or didn't) you love about this person?
Q7-2.   How do you feel about this person having such qualities?
Q7-3.   How did you try to hurt him (or her)?
Q7-4.   How did you cut yourself off from him?
Q7-5.   What else did you do in your relations with him?
Q7-6.   Can you, by putting yourself in his shoes, understand how and why he would have these qualities?
Q7-7.   How did you try to protect him and help him?
Q7-8.   How did you feel whenever you failed at this?
Q7-9.   Can you forgive yourself for having failed?

If you feel like crying, by all means do so. Let it out. You'll feel better. This particularly applies to men.

This exercise can be done as often as desired. Resolving your own grief can do as much for you, or more for you, than almost any other single thing.

End of Session

Work Session #8

Topic: Hate

Hate is the response to frustration.

For this session, the tracker should obtain a sharp knife and hold it in his hands, fondle it, or do anything he wants to with it throughout the session.

Q8- 1.   Have you ever wanted to kill anyone?
Q8- 2.   Why didn't you?
Q8- 3.   Look closely now at the ideal or ideals you give as a basis for not killing. Why do you accept them?
Q8- 4.   Quite apart from the matter of killing, consider this ideal. What is its opposite?

Q8- 5. What would happen if you adopted this opposite as a basis for living?

Q8- 6. Imagine now a situation in which you have been told to do something (other than killing) that violates this ideal. How would you feel?

Q8- 7. Let us suppose that despite your objections you are *forced* to do this thing. How would you feel now?

Q8- 8. Take this feeling and intensify it as much as you can.

Q8- 9. Now reduce the intensity until you can barely perceive it.

Q8-10. Now intensify it again, not to maximum, but to a high level, and observe your sensations in: head, back of neck, chest, stomach, throat, tongue, lips, teeth, gums, hands, buttocks, genitals, feet, knees, thumbs, shoulders.

Q8-11. Now eliminate this feeling. Have you ever felt this way about Mother?

Q8-12. Have you ever felt this way about Father?

Q8-13. Think of a situation in which you did (or might) feel this way. Visualize it in detail.

Q8-14. What did you do to get your own way in spite of everything?

End of Session

Work Session #9

Topic: Fear

Fear is a way of mobilizing our whole being to meet an emergency. It is useful to meet an emergency with all the power at our disposal.

In dealing with an emergency, we operate as though survival itself were at stake; in some cases it *is* at stake. As a result, we do *anything* that will help us control the emergency. This, again, is a healthy response in a well-designed organism.

But the things we do to deal with an emergency that WORK thereafter have a GREEN LIGHT character; we tend to favor acting that way in any future situation, especially if there is any degree of risk involved or if we are in any way disturbed or upset. This is not good, because the thing that happened to work in an emergency might be very inappropriate in some other situation.

Conversely, the things we do in an emergency that FAIL thereafter have a RED TAB character; we tend to *avoid-at-all-costs* doing things that way again, no matter what the situation. Again, this is not good, because such a RED TAB action might be the very thing that is needed in some later situation.

GREEN LIGHT for IT WORKS; RED TAB for IT FAILS: these were excellent survival mechanisms for a jungle life and still are where emergencies are common. But man invented civilization; when he did this he took responsibility for his own evolution; and he has created a very complicated world in which GREEN LIGHT and RED TAB often produce a great deal of harm.

Here, in germ, we have a basic explanation for irrational conduct, perhaps even for insanity itself. It is good to know this, because it helps us to understand why people do crazy things. We can assume, until proved otherwise, that they are operating under GREEN LIGHT or RED TAB patterns* that short out their rational minds. Hence, *we do not need to condemn them for their actions.* It would be much more intelligent to do what we can to help them to get rid of these patterns.

But while we are being intelligent and helpful, what about us? *We* have those patterns, too. What can be done to get rid of them in ourselves?

The answer, quite simply, is to *analyze* them, to re-evaluate them, to get them into proper perspective.

Each of us has had one or more emergencies in his life. Let's take a look, therefore, at each of these emergencies and see what we can learn from them.

Q9- 1. You're dying. What's the first thing to do?
Q9- 2. Your neck is broken. Above all, what mustn't be done?
Q9- 3. You're bleeding. Where's Mommy?
Q9- 4. It's all right now. You're safe. How do you feel?
Q9- 5. Feel this feeling of being safe, in turn, in: neck, stomach, throat, eyes, kidneys, buttocks, heart, breathing, stomach, throat.
Q9- 6. Review, now, your response to Q9-1, "You're dying." How did you feel?
Q9- 7. What did you think?
Q9- 8. Have you ever felt that way before?
Q9- 9. Have you ever thought that before?

---

*Such patterns are called "protodynes" in synergetics.

Q9-10. What do you think of this feeling now?

Q9-11. How do you feel about those thoughts now?

Q9-12. Consider now those thoughts. Observe, in turn, the following parts of your body for efforts in those parts. Think the thought and look for the effort at the same time, in each part of the body: heart, stomach, hands, shoulders, back, breathing, genitals, throat, mouth, eyes, nose, ears, face, feet, knees.

Q9-13. Repeat Q9-6 through Q9-12 for your response to Q9-2.

Q9-14. Ditto for Q9-3.

Q9-15. Define an emergency.

Q9-16. What are the major emergencies of your life? List all of them.

Q9-17. What are you going to do about it?

End of Session

Work Session #10

Topic: Love

Love is the most powerful of human emotions. Because we love someone or something, we will do almost anything. We will brave great dangers, endure untold hardship and privation, suffer misery and anguish and despair up to and beyond the breaking point, for someone or something we really love.

Conversely, we all want to *be* loved, too. Way down deep, we all want this very much, so much that we have to protect ourselves from people without scruples who might take advantage of us. Nothing hurts us as deeply as to give our love freely and to be laughed at.

We have a defense for this, too. We take our desire to be loved and call it "pride" or "vanity" or "self-esteem," and we take it out under this label and look at it and laugh at ourselves. It's a good defense. But deep down, we still want to be loved.

(And we still will do almost anything rather than to frankly acknowledge it to ourselves.)

Love often is paired off against intelligence, as if the two were opposites. This is understandable because often our heart tells us to do one thing and our head another. Usually we solve this by following our heart's lead. Whether this is wise or foolish is a matter for philosophers to debate.

But there is another solution: *use the two together.* Love without intelligence is often blind. Intelligence without love is often cold. Love needs reason to reach its goals more fully. Reason needs love to give meaning to the dry formulations of the intellect.

Together they make quite a team.

Q10- 1. What do you hate more than anything in the world?

Q10- 2. Underneath this hate there is a thwarted love. What do you love so much that makes you hate so much?

Q10- 3. Look at what you love. Have you given up?

Q10- 4. Think of a time when you were very lonely and depressed. Feel the loneliness, the misery, the despair.

Q10- 5. Intensify this lonely feeling and *feel* it in each of these parts of your body: legs, hair, teeth, arms, throat, heart, stomach, teeth, knees, elbows.

Q10- 6. You're ugly. What is the ugliest thing about you, not your body, about *you?*

Q10- 7. Take this ugliness and hate it. Despise it. Loathe it.

Q10- 8. Have you ever done this to yourself before?

Q10- 9. What are your thoughts now? Just report whatever comes to mind, without evaluation or censorship, for the next minute or so.

Q10-10. What do you think about life? Just present your thoughts as they occur, without trying to organize them or get rid of inconsistencies.

Q10-11. What do you think about sex? Do the same as before.

Q10-12. What do you think about world conditions? Again, the same.

Q10-13. If you were going to punish yourself internally, how would you go about doing it?

Q10-14. Think now of a time when you rejected somebody else. *Feel* your rejection of him again.

Q10-15. Think now of a time when you *forgave* somebody else for a wrong done to you. *Feel* your forgiveness of him again.

Q10-16. Now take a look at your own ugliness again. Forgive yourself as you forgave somebody else. Do unto yourself as you have already done unto another. Do not apologize or excuse, forgive.

Q10-17. In a little while an idea will be presented to you for your consideration. Now take this forgive-yourself feeling and *apply it* (don't feel it there) to the following parts of your body: elbows, knees, teeth, stomach, heart, throat, arms, teeth, hair, legs.

Q10-18. Is there any other part of your body that needs forgiving? Take a minute to do a thorough search, and if you find a part, forgive it as above.

Q10-19. What action have you done in the past of which you feel most ashamed?

Q10-20. Here is the idea mentioned earlier: you, alone, have the power to forgive yourself for anything you have done.
What do you think of this idea?

End of Session

Synergy Session #4

Q-1.   Review Work Sessions #7-10. Search for additional material that may have only partly emerged previously.

Q-2.   Carry out BAM Tracking (Analytical Procedure) on Responsibility.

Session Review

Work Session #11

Topic: Why Change?

For this session, simply repeat Work Session #1. Answer queries *fresh,* as if you had not done it before.

The reason for this is that if you have come this far, and *really* worked, your perspective has probably changed. Doing the first session over will throw those queries into an entirely different light. It will also give you a perspective of Operation Traverse as a whole.

End of Session

Work Session #12

Topic: Freedom

This is the last session of the series. But this is not an ending, it is only the beginning.

There is a new life for you awaiting. It is a life that will be full, free, and *fun*. There are so many things to do, so many things to experience, that one lifetime simply isn't enough. There must be something more after this is over; but it probably is so far different from this life-experience that we cannot even conceive it.

One thing I, personally, am sure of. There is no Hell. That's simply a projection of some theologian's guilt complex, a device to control, probably with a GREEN LIGHT or a RED TAB tacked on somewhere. The only Hell there is is that which we make for ourselves here and now.

Q12- 1. What is your evaluation of what you have achieved so far, using this Operation Traverse series?

Q12- 2. Name four things that you have always wanted to do, but which up to now you have never felt able to do before.

Q12- 3. Name a thing you have always wanted to do and *still* feel unable to do.

Q12- 4. Now look at this "unable" feeling. Feel it in each of these parts of your body: throat, stomach, heart, face, breathing.

Q12- 5. Now get up and pace the floor, back and forth. As you do this, feel in turn this unable feeling in the following movements:
— The swinging of your arms.
— The looking of your eyes as you move your gaze from object to object to object.
— The listening of your ears as you move your attention from sound to sound.
— The shift of balance you make just before you lift the rear foot up to put it ahead.
— The tension in your back as you turn.

Q12- 6. Now sit down and *feel exhausted*. You've never been so tired in all your life. Try, in turn, the following motions *without actually moving,* checking each by this *exhausted* feeling:

— Raise your right arm.
— Lift your left leg.
— Turn your head to the right.
— Actually close your eyes, and then try to open them but check the opening with this exhausted feeling.
— Now actually lie down on the floor or on a couch and relax a moment. Then try, without moving, to get up, but check it with this exhausted feeling.

Q12- 7. Repeat Q12-5, but this time instead of using the "feel unable" feeling to check movements, use a "don't want to but being forced to" feeling to resist movements. As you do this, compare the efforts in the two cases.

Q12- 8. Repeat Q12-6, but this time instead of using the "feel exhausted" feeling to prevent movements, use a "must but can't help myself" feeling to fight a struggle with yourself before finally giving in and *doing* the movements.

Q12- 9. What is freedom? Discuss this in any way you want to, giving your ideas for a few minutes.

Q12-10. What is the difference, if any, between freedom and self-determinism?

Q12-11. Now close your eyes and imagine yourself climbing a mountain. Feel the wind on your face, as if you were actually there. Make the mental and physical effort to move your arms and legs without actually moving them. Go through, mentally, the motions of:

— Jumping across a deep chasm six feet wide.
— Crawling on hands and knees through a short tunnel.
— Walking along a six-inch wide ledge with a cliff to your right and on your left a sharp drop-off.
— Walking for hours and hours up a steep grade with rocks and fallen trees and other obstacles of broken country in the way at nearly every step.
— Climbing by handholds and footholds up a short, steep stretch of solid rock.
— Reaching the summit and looking at the magnificent view in all directions.

Q12-12. Open your eyes. What is an obstacle?

Q12-13. Now close your eyes again and make the mental effort to run as fast as you can.

Q12-14. Open your eyes. What is an effort?

Q12-15. Close your eyes again and make the mental effort (feel it in your muscles) to slug somebody as hard as you can with your right fist.

Q12-16. Now make the mental effort to kick somebody who needs kicking just as hard as you can.

Q12-17. Open your eyes. What is the difference between living and just existing?

Q12-18. *Feel* this difference in the following parts of your body: head, arms, breathing, heart, stomach, back, genitals, breasts, throat, hands, shoulders, legs, feet, face.

Q12-19. Get up and walk across the room. Again, *feel* the difference in turn, in the following movements:

— Rise up on toes of foot.
— Plant of foot on surface.
— Swing of leg forward.
— Motion of shoulders.
— Locking of knee.
— Overall balance.

Q12-20. What is poise?

End of Session

Synergy Session #5

Q-1. Review Work Sessions #11-12. Search for additional material that may have only partly emerged previously.

Q-2. Read chapter 8.

Q-3. Select a present-time situation, person, group, or domain of activity. Do a Hyperception Sweep (apply, in turn, Tuning, Panview, Thrillbeam, and Clearlook).

Session Review

Protodyne Reduction Procedure

In chapter 9, protodynes were discussed. These are *Identic Mode patterns* that, when activated, operate an individual like a puppet on strings.

During a lifetime, a person can accumulate a large number of protodynes. He is unaware of them. He is unaware of the stimuli that set them off. Every time one of these unnoticed stimuli occurs, the protodyne runs its course—operating the individual without his knowledge or consent, causing him to do irrational things, often things he consciously struggles not to do but is powerless to prevent.

Protodynes are activated as a result of Self-Invalidations (S.I.s). They are an attempt by the organism to deal with situations that the conscious mind has decided it can't handle. They are an emergency mechanism, a regression to a more primitive level of function.

The important thing about this is that when the S.I. is cleared, the protodyne fades away, like an old soldier. The S.I., therefore, is the primary target.

However, when an S.I. is approached, the protodyne tends to be activated. This sometimes makes it difficult to clear an S.I. unless the *power* of the protodyne is reduced. Fortunately, this can be done without contacting the protodyne directly or in depth.

Protodyne Reduction Procedure enables the tracker to do this. It consists of seven query sequences, each focused on the general area of a major type of protodyne.

It is recommended that the tracker go through Protodyne Reduction Procedure at least once. Later, if he encounters difficulty with an S.I., one or more of the PRP sessions may be helpful.

Session #1—Need

Q- 1.   Recall or imagine a situation in which you needed someone.

Q- 2.   Recall or imagine a situation in which someone needed you.

Q- 3.   Imagine yourself hating the someone of Q-2. Be really ferocious.

Q- 4.   Do a body scan.

Q- 5.   Consider again the situation of Q-1. Imagine the someone being completely unresponsive to your need.

Q- 6.   Consider again the someone of Q-3. Imagine the someone completely unresponsive to your hate.

Q- 7.   Consider again the situation of Q-1. Imagine the someone ridiculing you for that need.

Q- 8.   Phase shift on your response to Q-7.

Q- 9.   State your BAM about authority.

Q-10.   Why do you accept this BAM?

Q-11.   What other BAM might you equally well adopt?

Q-12.   Recall or imagine a situation in which you were completely independent of anybody.

Q-13.   What needs (in Q-12) were you especially glad to be rid of?

Q-14.   Recall or imagine another situation in which you needed someone.

Q-15.   What ways did or might you use to control that someone?

Q-16.   Consider your mother. What needs did she fulfill for you as a child?

Q-17.   Do you still have those needs now? If so, why?

Q-18.   What needs did your father fulfill for you as a child?

Q-19.   Do you still have those needs now? If so, why?

Session Review

Session #2—Control

Q- 1.   Recall or imagine a situation in which you were controlled by someone.

Q- 2.   What factors and people control you in present time?

Q- 3.   Float. (Close your eyes and let images, ideas, feelings, memories, etc. come to mind freely. Repress nothing. Simply observe what happens.)

Q- 4.   Recall or imagine a situation in which someone treated you unjustly.

Q- 5.   Imagine yourself getting revenge.
Q- 6.   Phase shift on your response to Q-5.
Q- 7.   State your BAM about freedom.
Q- 8.   Why do you accept this BAM?
Q- 9.   What other BAM might you equally well adopt?
Q-10.   Recall or imagine a situation in which you controlled someone.
Q-11.   What checks did you impose on your control?
Q-12.   Why? (Look for values)
Q-13.   How else might you fulfill these values?
Q-14.   Recall or imagine a situation in which you tried to control yourself and failed.
Q-15.   Consider again your response to Q-1. How did you resist or rebel against that control?
Q-16.   Consider again your response to Q-14. What caused you to fail?
Q-17.   Compare this cause or causes with your response to Q-15.
Q-18.   How did your mother control you as a child?
Q-19.   How did you resist or rebel against that control?
Q-20.   How did your father control you as a child?
Q-21.   How did you resist or rebel against that control?

Session Review

Synergy Session #6

Q-1.   Review PRP Sessions #1-2. Search for additional material that may have only partly emerged previously.
Q-2.   Read chapter 10.

Session #3—Hate

Q- 1.   Make a list of the things you hate.
Q- 2.   For each item on the list state rationally why you hate it.
Q- 3.   Close your eyes, and go over the list and *feel* the hate. What would you like to *do* in each case?
Q- 4.   Do a body scan, relating body sensations to your feelings of Q-3.
Q- 5.   Review your response to Q-3. What checks on action did or might you impose?

Q- 6.   Select one of these checks. What *value* did it serve?

Q- 7.   How else might you fulfill this value?

Q- 8.   You're a very clever person. How did or might you get around this check and express hate anyway?

Q- 9.   Repeat Q-6 through Q-8 for another check.

Q-10.   State your BAM about cruelty.

Q-11.   Recall or imagine a situation in which someone hated you. How did or might you respond?

Q-12.   Compare the way you responded with your list of Q-1. Does your response fit any of the items on the list?

Q-13.   (Optional) Repeat Q-11 and Q-12.

Q-14.   What other BAM about cruelty might you equally well adopt?

Q-15.   Recall or imagine a situation in which you hated someone. Make the reason for hating a good one and *feel* the hate.

Q-16.   What does this feeling make you want to *do*?

Q-17.   What checks on doing these things do or might you impose?

Q-18   What value does this serve?

Q-19   How else might you fulfill this value?

Q-20.   You're a very clever person. You know what query to ask now!

Q-21.   In what way or ways did you hate your mother as a child?

Q-22.   Your father?

Q-23.   For each: how would you have *liked* to express this hate?

Q-24.   How did you *actually* express it?

Session Review

Session #4—Fear

Q- 1.   Recall or imagine a situation in which you are overwhelmed by panic.

Q- 2.   Do a body scan and *feel* the panic in various parts of your body.

Q- 3.   Consider again your response to Q-1. Look for things you feel you *must* do but are prevented from doing.

Q- 4.   Select one of these things. What value compels you to do this thing?

Q- 5.   How else might you fulfill this value?

Q- 6.   What value prevents this thing from being done by you?

Q- 7.   How else might you fulfill this value?

Q- 8.   Do a prime shift on the values of Q-4 and Q-6.

Q- 9.   Repeat Q-1 through Q-8.

Q-10.   Make a list of things you are afraid of.

Q-11.   Select one of these things. State your BAM about it.

Q-12.   Why do you accept this BAM?

Q-13.   What other BAM might you equally well adopt?

Q-14.   Recall or imagine a situation in which you are afraid of doing something harmful to yourself.

Q-15.   Imagine yourself doing it.

Q-16.   What value would be promoted by doing this thing?

Q-17.   What prevents you from fulfilling this value? (Look for a second value)

Q-18.   How else might you fulfill this second value?

Q-19.   State your BAM about courage.

Q-20   Why do you accept this BAM?

Q-21.   What other BAM might you equally well adopt?

Q-22.   In what ways do you bind yourself because of hidden hate?

Session Review

Synergy Session #7

Q-1.   Review PRP Sessions #3-4. Search for additional material that may have only partly emerged previously.

Q-2.   Carry out Analytical Procedure on a topic of your choice.

Session #5—Grief

Q- 1.   Recall a situation in which you lost something you highly valued. Feel the feelings you felt.

Q- 2.   Phase shift on your response in Q-1.

Q- 3.   What did you do or want to do to deny or reverse the loss?

Q- 4.   Did you feel anger toward the lost object for depriving you of a cherished value?

Q- 5.   Make a list of things you cherish most highly.

Q- 6.   Select one of these things. Imagine yourself losing it.

Q- 7.   What value does this thing serve?

Q- 8.   How else might you fulfill this value?

Q- 9.   Float.

Q-10.   Consider again your response to Q-1. What might you have done to prevent the loss?

Q-11.   Select one of the things on your list of Q-6. Is it a substitute for a thing you lost in the past?

Q-12.   Consider again your responses to Q-3 and Q-10. Look for ways in which these efforts to deny, reverse, or prevent the loss may be operating in present time.

Q-13.   Do a body scan, relating body sensations to the efforts of Q-12.

Q-14.   What was the first great loss of your life? (Comment: A loss that appears trivial to an adult may be catastrophic to a child.)

Q-15.   Feel the feelings you felt.

Q-16.   What did you do or want to do to deny, reverse, or prevent the loss?

Q-17.   What did you do or want to do at the instant you accepted the loss as irrevocable?

Q-18.   State your BAM about love.

Q-19.   Why do you accept this BAM?

Q-20.   What other BAM might you equally well adopt?

Q-21.   In what ways do you needlessly restrict yourself because of fear?

Session Review

Session #6—Pain

Q- 1.   Make a list of things you find painful or unpleasant.

Q- 2.   Select one of these things and recall or imagine a situation in which this occurs.

Q- 3.   Phase shift on your response to Q-2. What did or might you do to avoid, ignore, fight, control, and submit to the pain? Try to find an example of each.

Q- 4.   Do a body scan, relating body sensations to the efforts of Q-3.

Q- 5.   Make a list of things you find pleasant and enjoyable.

Q- 6.   For each thing on the list, think of its opposite (producing pain or unpleasantness for you).

Q- 7.   Select one thing on the list of Q-5, and recall or imagine a situation in which the occurrence of pleasure or enjoyment was thwarted, no matter what you do.

Q- 8. Phase shift on your response to Q-7. What was your response to the pleasure-thwart?

Q- 9. What else might you do to overcome the thwart?

Q-10. Consider the thing giving pleasure in Q-5. Do a prime shift.

Q-11. Select one of the pairs of opposites in which you used the pleasure to avoid the pain.

Q-12. Select another thing on your list of Q-1 and recall or imagine a situation in which you amplified the unpleasantness as a means of getting sympathy.

Q-13. Imagine your efforts to get sympathy thwarted and imagine ways to overcome the thwart.

Q-14. Now imagine a situation in which someone used this way of overcoming the thwart on you. What is your response?

Q-15. State your BAM about kindness.

Q-16. Why do you accept this BAM?

Q-17. What other BAM might you equally well adopt?

Q-18. In what areas have you arrested your own development because of efforts to deny, reverse, or prevent a past loss?

Session Review

Session #7—Guilt

Q- 1. Make a list of things you regard as morally wrong.

Q- 2. Select one of these things and recall or imagine a situation in which you did this thing.

Q- 3. Imagine yourself getting caught. How do you respond?

Q- 4. What values are promoted by this response?

Q- 5. How else might you promote these values?

Q- 6. What reasons might you give to justify your "wrong" action of Q-2?

Q- 7. Probe deeply now. What's the real reason?

Q- 8. Recall or imagine a situation in which you did something of which you were later ashamed.

Q- 9. Feel the shame and do a body scan.

Q-10. What values were protected by "feeling ashamed"?

Q-11. How else might you promote these values?

Q-12. Recall or imagine a situation in which you caught someone else doing something wrong.

Q-13.   Imagine yourself meting out stern justice. What would you do?

Q-14.   Make a list of the things you would secretly like to do but don't because society might disapprove.

Q-15.   Select one of these things and imagine yourself doing it. Have a ball!

Q-16.   Now consider the stern justice you administered in Q-13. Consider the action only, quite apart from the "justice," and imagine a situation in which you do this for the pleasure it gives you.

Q-17.   Now imagine a society in which this action is regarded as wrong. How would you respond?

Q-18.   Make a list of things you really don't like to do, but do because society expects it of you.

Q-19.   State your BAM about responsibility.

Q-20.   Why do you accept this BAM?

Q-21.   What other BAM might you equally well adopt?

Q-22.   In what areas have you stopped or reduced function in order to avoid possible pain?

Session Review

Synergy Session #8

Q-1.   Review PRP Sessions #5-7. Search for additional material that may have only partly emerged previously.

Q-2.   Sweep the Mode Ladder looking for present-time processes that may be going on at each rung.

Operation Milquetoast

Please read again the section on Self-Invalidations in chapter 9.
The purpose of Operation Milquetoast is to help the tracker contact his S.I.s and to reduce their intensity. Later, Creative Procedure enables him to systematically eliminate all of them.

It is vitally important that the tracker keep an S.I. Log. Each time he contacts an S.I., he *immediately* writes it down in his S.I. Log. It is also desirable to examine and write down the associated Prime Determinant at the same time. Later, using Creative Tracking, he will eliminate all the S.I.s noted in his S.I. Log.

Operation Milquetoast consists of five query sequences. Each is aimed at a situation type that often evokes S.I.s. The situations are: thwart, threat, loss, justice, and social invalidations.

Session #1—Thwart

Q- 1.   How do you feel when you are thwarted—prevented from doing something you want to do?

Q- 2.   How do you feel when made to do something you don't want to do?

Q- 3.   Recall or imagine an incident in which you felt this way (do for both Q-1 and Q-2).

Q- 4.   How do you feel when you get your way in spite of obstacles?

Q- 5.   Recall an incident in which this happened.

Q- 6.   Consider a specific thwart. What thwarts you? What are you unRAW to do or experience?

Q- 7.   What ability would you need to do or experience this thing?

Q- 8.   Imagine a situation in which you have this ability in abundance. Be superman and describe what you would do!

Q- 9.   Consider again your response to Q-6. Take a fresh look—anything else you are unRAW to do or experience?

Q-10.   What BAM do you hold about doing or experiencing this thing?

Q-11.   What other BAM might you equally well adopt?

Q-12.   In what way or ways did you *turn yourself off* because you were unRAW? Pinpoint the Self-Invalidation. Look for:

   a.   An agreement that restricts or usurps your freedom.
   b.   A judgment: "I am inadequate."
   c.   A self-rejection.
   d.   An acceptance of external control.
   e.   A self-control.
   f.   A VIP-bond—acceptance of the Value-Interest-Perspective of some group, class, ideology, etc.
   g.   A Noble Burden.
   h.   A Confusion of Identity.
   Note: Be sure to write this down in your S.I. Log.

Q-13.   What value do you hold about not doing or experiencing this thing?

Q-14.   What is the opposite of this value?

Q-15.   In what way or ways do you promote the opposite of this value?

Q-16.   Look now for the DETERMINATION—the covert, reactive way of getting your own way anyhow.

Q-17.   What do you think of all this?

235

Q-18.   Look again at the DETERMINATION. Why do you regard this as "bad"?

Q-19.   What other way might you fulfill the same goal?

Q-20.   Look again at the thing you are unRAW to do or experience. Why does this thwart you?

Q-21.   What would happen if you did do or experience this?

Q-22.   Why do you believe this? What data is this belief based upon? What interpretations did you make of these data? What other interpretations are possible?

Q-23.   Formulate a Synergic Self-Affirmation—adopt a synergic BAM in place of the Self-Invalidation.

Q-24.   What can you do to bring your responsibility into balance with your freedom and power? (This is called unity of RFP.)

Q-25.   What commitments and/or obligations do you wish to reconsider? How can you adopt these synergically?

Session Review

Session #2—Threat

Q- 1.   How do you feel when you are in physical danger? Gently recapture the feeling and look at the ways you respond. Make a list of them. Add to the list later if you think of other things.

Q- 2.   How do you feel when you are in social danger (risk of ridicule, exposure of something you are secretly ashamed of, irrevocable loss of status, prestige, affection, etc.)? Recapture the feeling gently and look at the ways you respond. Make a list of them. Add to the list later.

Q- 3.   Recall or imagine an incident in which you felt in physical danger. Try to pinpoint the major details exactly. Don't relive the incident; re-examine it objectively.

Q- 4.   Recall or imagine an incident in which you felt you were in social danger. Try to pinpoint the major details exactly, as in Q-3.

Q- 5.   Consider again the physical danger incident and ask yourself: What were you trying to prevent from happening? What were you unRAW to experience or do?

Q- 6.   What is courage? Give an example of a courageous act.

Q- 7.   What is cowardice? Give an example of a cowardly act.

Q- 8.   Consider the social danger incident and ask yourself: What were you trying to prevent from happening? What were you unRAW to experience or to do?

Q- 9.   Consider courage again. Think of another example. Have you ever tried to act that way?

Q-10.   (a) If yes, look at this pattern objectively. How might it impede your freedom of action?

(b) If no, look for a way you have rejected yourself. Ask yourself why.

Q-11.   Consider again the physical danger incident and the "preventer" you installed. In what way or ways did you turn yourself off? Look for:

a. An agreement that restricts or usurps your freedom.
b. A judgment: "I am inadequate."
c. A self-rejection.
d. An acceptance of external control.
e. A self-control.
f. A VIP-bond—acceptance of the Value-Interest-Perspective of some group, class, ideology, etc.
g. A Noble Burden.
h. A Confusion of Identity.

Q-12.   Now consider again the social danger incident and the "preventer" you installed. In what way or ways did you turn yourself off? Look for one or more of the items listed in Q-11. Be sure to note these S.I.s in your S.I. Log.

Q-13.   Consider cowardice again. Think of another example. Have you ever acted, or felt like acting, this way?

Q-14.   (a) If yes, did you punish yourself or try to atone for acting or feeling this way? Look at the punish/atone pattern objectively. How might this impede your freedom of action?

(b) If no, how would you feel about somebody else acting this way? Look at this feeling objectively. Does it lead you toward any agreements or intentions that might restrict your freedom?

Q-15.   Consider again the physical danger S.I. (Q-11). Look now for a Determination—a pattern to get your own way in spite of it all.

Q-16.   Consider again the social danger S.I. (Q-12). Look for the Determination.

APPENDIX

Q-17.   Consider the physical danger Determination. What other way might you get your own way in spite of it all?

Q-18.   Consider the social danger Determination. What other way might you get your own way in spite of it all?

Q-19.   Look again at the physical danger and what you were unRAW to experience or to do. What would happen if you did do or experience this?

Q-20.   Why do you believe this? What data is this belief based upon? What interpretations did you make of these data? What other interpretations are possible?

Q-21.   Look again at the social danger and what you were unRAW to experience or to do. What would happen if you did do or experience this?

Q-22.   Why do you believe this? What data is this belief based upon? What interpretations did you make of these data? What other interpretations are possible?

Q-23.   For the social danger S.I., formulate a Synergic Self-Affirmation—adopt a synergic BAM in place of the Self-Invalidation.

Q-24.   Apply unity of RFP. What can you do to bring your responsibility into balance with your freedom and power?

Q-25.   What commitments and/or obligations do you wish to reconsider? How can you adopt these synergically?

Q-26.   Repeat Q-23 through Q-25 for the physical danger S.I.

Session Review

Synergy Session #9

Q-1.   Review OM Sessions #1-2. Search for additional material that may have only partly emerged previously.

Q-2.   Read chapter 16.

Session #3—Loss

Note: If you feel like crying during this session, by all means do so. A grief discharge can be most beneficial.

This is an unusually long session, and you may wish to do it in two parts. A suggested break point is after Q-19. Be sure to do a Session Review there if you do decide to break.

Q- 1.   How do you feel when you have lost something you love? Gently recapture the feeling and look at the ways you respond. Make a list. Add to it later as you think of other ways.

Q- 2.   Imagine or recall a situation in which you are called upon to sacrifice yourself or something you value. Describe the way you would or did respond. Make a list.

Q- 3.   Imagine or recall a situation in which you want or expect someone to sacrifice himself or something he valued. Assuming he is reluctant, devise ways to get him to do what you want.

Q- 4.   Recall a time when you lost someone you valued, either through death or change of situation. How did you feel? Recapture the feeling gently and note your responses.

Q- 5.   Make a list of physical things you need or would prefer not to do without. For each, ask yourself why you need this.

Q- 6.   Make a list of social things you need or would prefer not to do without. For each, ask yourself why you need this.

Q- 7.   Select one of the items of Q-5. What is your emotional response to lacking it or having it taken away?

Q- 8.   Select one of the items of Q-6. What is your emotional response to lacking it or having it taken away?

Q- 9.   Consider the item of Q-7. To get or keep this item, what did you or would you agree to? Pinpoint the S.I. and write it down in your S.I. Log.

Q-10.   Look at this agreement. In what way or ways does or might this restrict your freedom of action?

Q-11.   Formulate a Synergic Self-Affirmation—adopt a synergic BAM in place of the S.I.

Q-12.   Apply unity of RFP. What can you do to bring your responsibility into balance with your freedom and power?

Q-13.   What commitments and/or obligations do you wish to reconsider? How can you adopt these synergically?

Q-14.   Consider the item of Q-8. To get or keep this item, what did you or would you agree to? Pinpoint the S.I. and write it down in your S.I. Log.

Q-15.   Look at this agreement. In what way or ways does or might this restrict your freedom of action?

Q-16.   What way or ways did or might you get your own way anyhow? Look for a Determination.

Q-17.   Formulate a Synergic Self-Affirmation—adopt a synergic

BAM in place of the S.I.

Q-18.   Apply unity of RFP. What can you do to bring your responsibility into balance with your freedom and power?

Q-19.   What commitments and/or obligations do you wish to reconsider? How can you adopt these synergically?

Q-20.   Recall or imagine a situation in which you are forced to give up something you regard as absolutely necessary. How would (or did) you respond?

Q-21.   Look closely at your response. Specifically, what were you unRAW to experience or to do?

Q-22.   Look again at your response. In what way might or did you turn yourself off as a way of coping with the situation?

Q-23.   In what way might or did you get your own way in spite of everything?

Q-24.   In what way or ways are you ready to help others? Make a list.

Q-25.   Select one item on this list. What restriction would you accept to get somebody to help you this way?

Q-26.   Have you ever tried to impose this restriction on someone to have him do as you wish?

Q-27.   In what ways are you unwilling to help others? Make a list.

Q-28.   Select one item on this list and imagine you are forced to give this help. How would you respond?

Q-29.   Look at this response. In what way or ways might this restrict your freedom of action?

Q-30.   In what way or ways can you be forced to do something against your will? Make a list.

Q-31.   Select one item on this list. What would you do to get even if you could?

Q-32.   Look at this way of getting even. In what way or ways might this restrict your freedom of action?

Q-33.   Make a list of ways in which people should be required to do things for their own good.

Q-34.   Select one of these ways. Imagine (or recall) it being done to you. How might or did your respond?

Q-35.   What would or did you give up in accepting this discipline?

Q-36.   Look at your response. In what way might or did you turn yourself off when you accepted this discipline? Pinpoint the S.I. and write it down in your S.I. Log.

Q-37.   In what way might or did you get your way in spite of everything?

Q-38.   Formulate a Synergic Self-Affirmation—adopt a synergic BAM to replace the S.I.

Q-39.   Apply unity of RFP. What can you do to bring your freedom into balance with your responsibility and power?

Q-40.   What commitments and/or obligations do you wish to reconsider? How can you adopt these synergically?

Session Review

Synergy Session #10

Q-1.   Review OM Session #3. Search for additional material that may have only partly emerged previously.

Q-2.   Review the Tracker's Guide. Evaluate your work effort in terms of each policy.

Session #4—Justice

Q- 1.   Recall a time when you were treated unfairly and unjustly. How did you feel?

Q- 2.   What did you want to do to get justice? In what way or ways might you have made the injury worse than it actually was?

Q- 3.   In what way or ways may you have turned yourself off in order to get justice? Pinpoint the S.I. and write it down in your S.I. Log.

Q- 4.   How did you get your own way in spite of everything? Pinpoint the Determination.

Q- 5.   How else might you get your own way in a more rational and synergic manner?

Q- 6.   Formulate a Synergic Self-Affirmation—adopt a synergic BAM in place of the S.I.

Q- 7.   Apply unity of RFP—what can you do to bring your power into balance with your freedom and responsibility?

Q- 8.   What commitment and/or obligations do you wish to reconsider? How can you adopt these synergically?

Q- 9.   Recall a time when you accused someone else of doing something he didn't do. What were you trying to gain?

Q-10.   How else might you have got what you wanted?

Q-11.   In what way or ways might you have turned yourself off as a result of this effort? Pinpoint the S.I. and write it down in your S.I. Log.

Q-12.   How did you get your own way in spite of everything? Pinpoint the Determination.

Q-13.   How else might you have gotten your own way more rationally and synergically?

Q-14.   Repeat Q-6 through Q-8 for the S.I. of Q-11.

Q-15.   Recall a time when you did something of which you were later ashamed.

Q-16.   Who might you have been trying to please? Or hurt?

Q-17.   In what way or ways might you have turned yourself off as a result of this effort to please or hurt? Pinpoint the S.I. and write it down in your S.I. Log.

Q-18.   How did you get your own way in spite of everything? Find the Determination.

Q-19.   Repeat Q-6 through Q-8 for the S.I. of Q-17.

Q-20.   Recall a time when you deliberately did something you "knew" was wrong, i.e., something that you, yourself, accepted as wrong.

Q-21.   In what way or ways did you seek to "place the blame" on somebody else for your action?

Q-22.   In what way or ways did you turn yourself off as a result of this effort? Pinpoint the S.I. and write it down in your S.I. Log.

Q-23.   In what way or ways did you secretly try to atone for your action?

Q-24.   In what way or ways did you secretly try to get caught?

Q-25.   How did you get your own way in spite of everything? Pinpoint the Determination.

Q-26.   How else might you have gotten your own way more rationally and synergically?

Q-27.   Repeat Q-6 through Q-8 for the S.I. of Q-22.

Session Review

Session #5

Social Invalidations

It should be emphasized that the average person does not "cause" his own Self-Invalidations. An S.I. occurs as one link in a causal chain, which includes the actions and communications of other people. In particular, there are two types of external dysergy that add greatly to every person's dysergy load. These are the Impedances of Others (abbreviated "impots"), and sociodynes.

Ideally, everyone should clear his own impedances; but this is usually difficult to arrange. The syngeneer himself cannot clear an impot, only the other person can; and if he won't, you cannot convince a man against his will that he has one.

This means that the syngeneer must somehow deal with the impot. There are three aspects to impot management:

1. Neutralization. While it is impossible for one person to clear the impedance of another, *he can eliminate its dysergic effect on him.* This is called *neutralization.*

Neutralization can be done by tracking one's reaction to a Determinant—for example, a General Esteem Loss point—and then substituting a synergic response for the reaction. This doesn't clear the impot or the difficulties it causes, but the tracker can still accomplish far more if he is himself synergic.

2. Analysis. By observation and quiet queries, the syngeneer may gain a considerable degree of understanding of the impot. Joe's nasty temper may result from the fact that as a child his parents ignored him when he needed them, unless he strongly expressed anger. This implies an underlying S.I. that may have occurred when Joe encountered a situation he could not handle. Understanding Joe's impedance raises the syngeneer's esteem for Joe: it also enables him to predict Joe's probable behavior and to take preventive measures.

3. Finessing the impot. By this is meant the adoption of policies and practices that avoid stimulating the impot, that isolate or divert its dysergic effects into harmless channels if it is stimulated, that enable the syngeneer to achieve his goals in spite of such dysergy, etc. Finessing is an art. Some syngeneers become quite good at it.

Neutralization, analysis, and finessing are also very useful in dealing with sociodynes. Here, the syngeneer is concerned with the Identic and Reactive *patterns-of-expected-conduct,* which a sociodyne imposes upon him by virtue of his memberships, statuses, and roles.

Sociodynes produce Social Invalidations of the individual. These, in turn, (Identic Mode!) are echoed by individual Self-Invalidations. Thus, if a person is rejected by a group, he rejects himself.

Each of us is a member of a variety of groups, using "group" in its broadest sense. These include:

Family	Work Group
Ancestry	Labor Unions
Race	Business or Professional
Socioeconomic class	Organizations
Religion	Nation
Neighborhood	Etc.

For each of these groups, the individual has a Group Member Consciousness, distinct from, though interacting with, his Individual Consciousness. This distinction should be borne in mind in the following. Shift back and forth as appropriate; this is a very enlightening exercise!

Q- 1. Select a group (from the above list as expanded for your case). What is your status in the group? What are your roles in the group?

Q- 2. Recall or imagine an incident in which you did (or failed to do) something the group disapproved of (or would disapprove of if it knew).

Q- 3. Select a second group. What is your status? What are your roles?

Q- 4. Repeat for a third group.

Q- 5. Review your response to Q-2, adding further details as they occur. How did you *feel* about this disapproval?

Q- 6. Close your eyes and *feel* this feeling in various parts of your body.

Q- 7. Repeat Q-2 for the third group.

Q- 8. Consider now the second group. Recall or imagine an incident in which you were unjustly treated because you were a member of the group.

Q- 9. Close your eyes and do nothing for a few moments, observing and noting whatever happens.

Q-10. Consider again your response to Q-2. In what way or ways did you *stop developing* as a result of group disapproval (if it knew) or of your hiding the facts (if it didn't)? Pinpoint the Self-Invalidation and write it in your S.I. Log.

Q-11. Repeat for your response to Q-7.

Q-12. Consider now the second group. Recall or imagine an incident in which you failed or were inadequate in performing one of your roles.

Q-13.   In what way or ways did you stop developing as a result of this? Pinpoint the S.I. and add it to your S.I. Log.

Q-14.   Repeat Q-12 and Q-13 for the first group.

Q-15.   Consider now the third group. Recall or imagine an incident in which you disagreed strongly with the group consensus.

Q-16.   Repeat for the first group.

Q-17.   Consider now the second group. Imagine now that you are a member of a different group of this type (e.g., a different race), a group "lower in status." Imagine an incident in which this lower status restricted your freedom of action.

Q-18.   Repeat this for the third group.

Q-19.   Consider now your response to Q-15. In what way or ways did you stop developing as a result of this disagreement? Pinpoint the S.I. and enter it in your S.I. Log.

Q-20.   Repeat for your response to Q-16.

Session Review

Synergy Session #11

Q-1.   Review OM Sessions #4-5. Search for additional material that may have only partly emerged previously.

Q-2.   Read the sections on Creative Tracking and the Dysergy Converters in chapter 10.

Creative Procedure

Affectionately dedicated to Lorraine Cullen and Richard McMahan.

Please read again the sections on Self-Invalidations in chapter 9 and Creative Tracking and the Dysergy Converters in chapter 10.

The purpose of Creative Procedure is to enable the tracker to systematically eliminate all his S.I.s. To accomplish this, the following are emphasized:

1. Keep your S.I. Log up-to-date. Whenever you encounter another S.I., add it immediately to the Log. Be sure also to include the Prime Determinant.

2. Know Creative Tracking *well,* and apply it thoroughly, with total, objective self-honesty. In particular, *always* follow through with the Experiment, Develop, Apply operations of the FEDA Sequence.

3. The Dysergy Converters are powerful tools. Know them well and use them thoroughly.

4. Develop an ever-growing and ever-deepening awareness of the process of creative evolution, which flows inexorably onward at the roots of your being. This is nonverbal, although it helps to characterize it verbally. Focus on your uniqueness as an individual, on the uniqueness of each moment, on the infinite potential for developing new ideas, new ways of doing things, new and deeper understanding of yourself and your environment. Regard each session, each situation, as in some ways *unprecedented,* requiring a new approach. Be alert for emergents and rapidly explore their implications. Be ready to take singular action—action that is unexpected, unpredictable in terms of past categories. Remember the heurism: the human mind can transcend itself repeatedly.

5. There is a difference between a Self-Invalidation, a turn-off of creative evolution in an area vital to the individual, and a rational apprehension of limits. A child feels he can reach out and grasp the moon in the sky. He is not invalidated when he learns that he cannot.

6. Experience has shown that there often exist "Counter S.I.s" associated with the primary S.I.s described in chapter 9. Thus:

S.I.	*Counter S.I.*
Agreement	Disagreement
Judgment of Inadequacy	Judgment of Super-adequacy
Self-rejection	Self-adulation

S.I.	*Counter S.I.*
Acceptance of control	Reactive rejection of control
Self-control	Self-indulgence
V.I.P. bond	Reactive V.I.P. rejection
Noble burden	Freeloading on others
Confusion of identity	Reactive role rejection

The Counter S.I. also turns off creative evolution. Creative Tracking is still applicable as described, except that the Prime Determinant may also require reversal.

S.I.	*Counter S.I.*
UnRAW horn	OverRAW horn
Forcing horn	Restraining horn

7. Finally, Creative Tracking not only clears S.I.s; it also turns on the synergic mode. As the tracker develops and applies synergic abilities to replace his S.I.s, he inevitably finds himself operating in the synergic mode more and more of the time. There is usually no dramatic moment, no specific instant of time, when a tracker stabilizes. One day he simply notices that he *is* synergic most of the time. If he re-reads the description of the synergic mode given in chapter 1, he finds that it accurately describes his mode of function. This is pleasant to know, but he pays it only passing attention. He is having too much fun doing synergic things!

In the following series of Work Sessions, Creative Tracking is systematically applied to the major types of S.I. Some queries are included to help locate and clarify each S.I., if one has not previously been found of that type.

Don't forget to complete the FEDA Sequence! (The EDA phase replaces the Synergy Sessions in Creative Procedure.)

Session #1—Agreements

Note: If you have already located an Agreement S.I., jump to Q-9. The queries Q-1—Q-8 are designed to help you locate such S.I.s.

Q- 1.   What are some things that you tend to be disagreeable about?

Q- 2.   Do you like to argue? If so, why? If not, why not?

Q- 3.   Recall or imagine an incident in which you disagreed with someone. What happened just before the disagreement? What happened just after?

Q- 4.   How do you feel, emotionally, about this disagreement? Close your eyes and try to feel this feeling again in various parts of your body.

Q- 5.   Recall or imagine an incident in which you were forced to agree to something but really didn't. What happened just before the agreement? What happened just after?

Q- 6.   Close your eyes and do nothing for a few moments, just observing whatever comes to mind.

Q- 7.   Consider again the agreement of Q-5. Does it remind you of another agreement? If so, choose which agreement you want to work on and make a note of the other.

Q- 8.   Is the agreement still in effect in any way? If so, proceed. If not, repeat Q-1—Q-7 one more time. If there still is no agreement, end the session.

Q- 9.   In what way or ways does the agreement restrict your freedom of action? Does it compel you to do something you'd rather not do? Does it keep you from doing something you'd really like to do?

Q-10.   Why do you continue the agreement? What was it, or what is it, that you were or are unRAW to experience or to do?

Q-11.   Why were or are you unRAW? What might happen if you did experience or do this thing? Why is this bad?

Q-12.   Would it be possible to do or experience this thing to some degree if circumstances warranted it?

Q-13.   Consider again the restriction of Q-9. In what way or ways might you turn this to advantage?

Q-14.   Take a fresh look at this agreement and your response to queries Q-9—Q-13. What new thing might you learn from all this?

Q-15.   Regard the whole situation as an opportunity to evolve. In what way or ways might you change?

Q-16.   Close your eyes and do nothing for a few moments, just observing whatever comes to mind.

Q-17.   Reread the quotation from Henri Bergson (chapter 9). As you do so, focus on the unique emergents of the present situation.

Q-18.   Look again at the situation that forced you to make the agreement. Are any impedances of others (impots) or sociodynes involved? If not, jump to Q-23.

Q-19.   What was or is your response to the impot or sociodyne?

Q-20.   Is there any other way you might respond more synergically and still get what you want?

Q-21.   In what way or ways might you outsmart the impot or sociodyne?

Q-22.   Repeat Q-19—Q-21 for all impots and sociodynes located in Q-18.

Q-23.   Look again at the agreement. In what way or ways did you get your own way in spite of everything? (This is the Determination.)

Q-24.   What ability or abilities did you develop as a result of the agreement? (This is the Compensation.)

Q-25.   SESSION REVIEW. Review the high points of your response to the previous queries.

Q-26.   Define an ability you would like to have or develop further if you already have it.

Q-27.   Analyze this ability into simple steps or mental operations.

Q-28.   Think of a test situation in which you might try out this ability.

End of Session

At some convenient time in the near future, complete the FEDA sequence by:

1. *Experiment*—Try out the ability in the test situation.
2. *Develop*—Evaluate the experience and modify the ability as indicated.
3. *Apply*—Repeat *Experiment* and *Develop* steps until the ability becomes established, so you can use it any time.

Session #2—Judgments of Inadequacy (JINADS)

Note: If you have already located a JINAD, jump to Q-9.

Q- 1.   What are some of the things you are good at? Make a list.

Q- 2.   Recall or imagine an incident in which you succeeded at something by using one of these abilities.

Q- 3.   What are some of the things you are not good at? Consider only inadequacies that have put you at a real disadvantage.

Q- 4.   Recall or imagine an incident in which you did something that embarrassed you or made you feel ashamed.

Q- 5.   Consider Q-3 again and try to think of other inadequacies.

Q- 6.   Recall or imagine an incident in which you failed at something that was important to you. What happened just before the failure? What happened just after?

Q- 7.   Close your eyes and do nothing for a few moments, just observing whatever comes to mind.

Q- 8.   Consider again the failure of Q-6. Did you make a judgment of inadequacy in response to the failure? If so, proceed. If not, repeat Q-1—Q-7 one more time. If there is still no JINAD, end the session.

Q- 9.   Consider the JINAD. Define as exactly as you can just what this inadequacy is. What was the basis for your judgment? In what way or ways does the JINAD restrict your freedom of action?

Q-10.   What was it, or what is it, that you were (are) unRAW to do or experience?

Q-11.   Why? What would happen if you did do or experience this thing?

Q-12   Consider again the restriction of Q-9. Are there any ways in which you impose the restriction on yourself? If so, what would happen if you released yourself from this restriction?

Q-13.   Take a fresh look at the JINAD and your response to queries Q-9—Q-12. What can you learn from this?

Q-14.   Regard all this as an opportunity to evolve. How might you do things differently?

Q-15.   Close your eyes and do nothing for a few minutes, just observing whatever comes to mind.

Q-16.   Look again at the situation leading to the JINAD. Are any impots or sociodynes involved? If not, jump to Q-21.

Q-17.   What was (is) your response to the impot or sociodyne?

Q-18.   Is there a more synergic way you could respond?

Q-19.   How might you outsmart the impot or sociodyne?

Q-20.   Repeat Q-17—Q-19 for any other impots or sociodynes.

Q-21.   Look again at the JINAD. In what way or ways did you get your own way in spite of everything? (Determination)

Q-22.   What ability or abilities did you develop as a result of the JINAD? (Compensation)

Q-23.   Session Review.

Q-24.   Define an ability you would like to have or, if you already have it, to develop further.

Q-25.   Analyze this ability into simple steps or mental operations.

Q-26.   Think of a test situation in which you might try out this ability.

End of Session

At some convenient time in the near future, complete the FEDA sequence:

1. *Experiment*—Try out the ability in the test situation.
2. *Develop*—Evaluate the experience and improve the ability.
3. *Apply*—Repeat *Experiment* and *Develop* steps until the ability is a part of your repertoire of skills.

Session #3—Self-Rejections

In previous sessions, all queries have been written out to make it easy for you. By this time you have gained experience with Creative Tracking so that this is no longer necessary. In this and subsequent sessions, therefore, queries will be used solely to help you locate the particular kind of S.I. involved. The remainder of the session or the next session consists of carrying out the steps of Creative Tracking. These steps are briefly repeated below; memorize them by repeated sweeps. This will speed up your work effort.

Q- 1.   What trait would you most like to have? Make a list of desirable traits. Call this your Ego Ideal.

Q- 2.   What trait would you least like to have? Make a list of undesirable traits. Call this your Anti-Ideal.

Q- 3.   Select a trait from your Ego Ideal. Do you have this trait? If yes, jump to Q-11.

Q- 4.   If no, how do you feel about it? Recapture this feeling gently. Describe it in words.

Q- 5.   What have you done, or what would you do, to possess this trait?

Q- 6.   What is the opposite of this trait? What is the value, to you, of being this way (having the opposite of the trait)?

Q- 7.   In what way or ways have you turned yourself off because you lack the Ideal Trait? Pinpoint the S.I. and add it to your S.I. Log.

Q- 8.   How do you get your own way in spite of everything?

Q- 9.   How else might you get your own way more rationally and synergically?

Q-10.   Jump to Q-18.

Q-11.   Feel good about having this trait. Really admire yourself— have a ball!

Q-12.   What would be the value to you of *not* having this trait?

Q-13.   Who else do you know who has this trait? Who admires or admired you for having this trait?

Q-14.   In what way or ways does having this trait restrict your freedom of action?

Q-15.   In what way or ways might you have turned yourself off in order to have this trait? Pinpoint the S.I. and add it to your S.I. Log.

Q-16.   How did you get your own way in spite of everything?

Q-17.   How else might you get your own way more rationally and synergically?

Q-18.   Select a trait from your Anti-Ideal. Do you have this trait? If no, jump to Q-24.

Q-19.   If yes, how do you feel about it?

Q-20.   How might having this trait give or have given you control over somebody?

Q-21.   In what way or ways might you have turned yourself off in order to keep this control? Pinpoint the S.I. and add it to your S.I. Log.

Q-22.   How else might you get your own way without having to control somebody else?

Q-23.   SESSION REVIEW.

Q-24.   How might not having this trait endear you to somebody?

Q-25.   In what secret ways might you like to have this trait? Assume you can get away with anything!

Q-26.   In what way or ways might you have turned yourself off by rejecting this trait? Pinpoint the S.I. and add it to your S.I. Log.

Q-27.   How did you get your own way in spite of everything?

Q-28.   Pause for a moment and admire your own cleverness.

Q-29.   How else might you get your way, rationally and synergically, without having to please somebody?

Session Review

Creative Tracking

Formulate

Step One: Define the S.I. as clearly as you can.

Step Two: Analyze the horns of the Prime Determinant— the unRAW horn and the forcing horn.

Step Three: Apply one or more of the Dysergy Converters:
Information Source
Mutation Source
Unique Focus
Buckmaster
Chinese Waterhammer (if situation gets really difficult)

Step Four: Neutralize, Analyze, and Finesse all impots and sociodynes involved.

Step Five: Define a Synergic Ability (or more) and think of a test situation.

Experiment

Try out the ability in the test situation.

Develop

Evaluate the experience, and improve the ability as indicated.

Apply

Repeat *Experiment* and *Develop* steps until the ability has become part of your repertoire of skills.

Session #4—Acceptances of Control

Don't forget—Take Charge Procedure!

Q- 1.   Who and what are the people, organizations, etc. that control you? Make a list.

Q- 2.   Select one. How does it control you? Why do you accept that control?

Q- 3.   Repeat for another on your list.

Q- 4.   Recall or imagine a time when you needed help. What happened?

Q- 5.   How do you feel about helping others? Be honest.

Q- 6.   Recall or imagine a time when you were forced to do something you didn't want to. How did you respond?

Q- 7.   Recall or imagine a time when you were kept from doing something you wanted to do. How did you feel?

Q- 8.   Close your eyes and feel this feeling in various parts of your body. Then do nothing for a few moments, just observing what happens.

Q 9.   Review the list of Q-1 and add to it if you can. Then repeat Q-2 for a *different* "controller."

Session Review

Q-10.   Select one of the three "controllers" from the list you have looked at and start Creative Tracking.

Session #5—Self-Controls

Q- 1.   Make a list of "bad habits" or patterns you'd like to eliminate.

Q- 2.   Select one of these. Why do you regard it as "bad"?

Q- 3.   Recall or imagine an incident in which you condemned yourself for having this bad thing.

Q- 4.   Close your eyes, and apply this feeling of self-condemnation to various parts of your body—eyes, teeth, nose, lips, mouth, neck, throat, etc. (This is called "Body Scan.")

Q- 5.   What value does this habit have for you? (Pleasure, release from tension, etc.)

Q- 6.   What other way or ways might you fulfill this value?

Q- 7.   Repeat Q-2—Q-6 for a second bad habit or pattern.

Q- 8.   Define selfishness. Are you in any way selfish? If so, how do you feel about it?

Q- 9.   Define greed. Are you in any way greedy? If so, how do you feel about it?

Q-10.   Close your eyes and do nothing for a few moments, just observing whatever happens. (This is called "Float.")

Q-11.   Recall or imagine a situation in which you were deprived of something you needed or very much wanted. How did you feel?

Q-12.   Do a Body Scan applying this feeling.

Q-13.   Recall or imagine a situation in which you were taken advantage of.

Q-14.   Float.

Q-15.   Consider again your response to Q-3. Who might you have been trying to please?

Q-16.   Do a Body Scan on your response to Q-8.

Q-17.   Consider again your response to Q-11. How did or might you get what you needed or wanted, or a substitute?

Q-18.   Do a Body Scan on your response to Q-9.

Q-19.   Consider again your response to Q-3. In what way or ways might you have turned yourself off? Pinpoint the S.I. and add it to your S.I. Log.

Q-20.   Repeat for the second "bad habit."

Q-21.   Consider again your response to Q-11. In what way or ways might you have turned yourself off? Pinpoint the S.I. and add it to your S.I. Log.

Q-22.   Repeat for your response to Q-13.

Session Review

Q-23.   Start Creative Tracking

Session #6—Value-Interest-Perspective (VIP) Bonds

Q-1.   Make a list of the groups, organizations, etc. to which you belong, including race, religion, socioeconomic class, nationality, etc.

Q-2.   Select one of these. What are your goals for this group?

Q-3.   What do you perceive as the Value-Interest-Perspective of this group?

Q-4.   How does this VIP affect your freedom of action?
Q-5.   Repeat Q-3 and Q-4 for another group, organization, etc.
Q-6.   Select one of these and start Creative Tracking.

Session #7—Noble Burdens

Love is one of the strongest of human emotions and the most beautiful. There are many kinds of love—man-woman mutual love, parent for child love, child for parent love, love for friends, love for groups, etc. Sometimes love is blind, and often it has reactive components. Love can turn to hate, indeed, most hate is a response to love disappointed. On the other hand, hatred can be overcome by love.

Whenever we lose someone we love—by death, by absence, by a severing of relations—we are hurt. Even a stable can be hurt. This is the price we pay for the ability to love, and if we consider the price too high, we turn off the ability. This is the sublime vulnerability of being human. Each must decide how vulnerable he wants to be.

The love we have for others leads us to do things on their behalf. This is especially so after we lose them. We assume, unknowingly, a Noble Burden. It is a way of keeping a part of those we have loved and lost.

Q-1.   Recall a time when you lost someone or something you loved. How did you feel?
Q-2.   Close your eyes and do nothing for a few moments, just observing whatever happens.
Q-3.   Recall another time when you lost someone or something you loved. What did you think at the instant you knew what had happened?
Q-4.   Consider again the loss of Q-1. What did you *gain?* Assuming you did gain something, how did you feel about it, and what did you do to hide the gain?
Q-5.   Repeat Q-4 for the loss of Q-3.
Q-6.   Recall another time when you lost someone or something you loved. What burden did you take on to compensate for the loss?
Q-7.   Repeat for the losses of Q-1 and Q-3.
Q-8.   Select one of these Noble Burdens and start Creative Tracking.

Session #8—Confusions of Identity

Who are you? Are you your name? Of course not, yet this is how we characteristically answer the question.

Q-1.   Close your eyes and say aloud your name—in whole or in part—over and over again. As you do so, look at the *nonverbal* aspects of your being. Do this for a few minutes.

Q-2.   What does the question, "Who are you?" mean? To answer it, we characteristically give, besides our names:

Our membership in various groups, organizations, etc.
Our family, race, religion, economic class, nationality, etc.
The status we have in each of these groups, etc.
The roles we play in each.

Systematically respond to the question for each of the above. As you do so, focus on the ways you are *not* each of these things.

Q-3.   Close your eyes and look again at the nonverbal aspects of your being. Try to get the feeling of your unique individuality, as distinguished from your various identities. Do this for your way of perceiving, your way of thinking, your way of feeling, your way of acting, your way of communicating with others.

Q-4.   Select one of your identities, and examine ways in which it may restrict your unique individuality.

Q-5.   Repeat for two more identities.

Q-6.   Select one of these three in which a restriction has been noted and start Creative Tracking.

Session #9—Planning Session

In this session, review the highlights of your work in Creative Procedure thus far. Then review your S.I. Log, including Counter S.I.s. Organize a schedule for systematically applying Creative Tracking to each S.I. Then start the first session. Continue Creative Tracking until all S.I.s have been cleared.

The Stabilization Telopause

As a stable, operating in the synergic mode most of the time, you naturally have a basically different perspective from what you had before. So much is now possible that you never dared dream of doing before! So full and exciting and precious is each moment of life!

Naturally, you have new personal goals to achieve. But as a synergic being, you also naturally want to help others to stabilize—not to "cause them to be like you," but to enable them to achieve *their* own unique potentials, free of S.I.s.

And you also want to contact other stables!

When two stables meet, the effect is electric. Each instantly recognizes the other *as a stable!* And as they interact, there soon emerges a synergic rapport, a relationship of love and trust and mutual admiration, of synergic teamwork, that transcends all other forms of human relationship. When two stables join minds in Totaltalk, the flow of communication is so fast and the empathy is so complete, that a new entity emerges—sometimes called Homo Gestalt. The mind of each is home to the mind of the other.

To facilitate this, and for other purposes, the Stabilization Telopause has been created. For further details, contact the Synergetic Society.

It sometimes happens that a mistake is made—that a person believes he has stabilized in the synergic mode when actually he has not. The Stabilization Telopause includes measures for handling this minor problem and for enabling true stabilization to occur.

The stables have other plans as well—to achieve modes of being beyond that of a stable—and to do their share toward resolving the problems of humankind.

NOTES AND REFERENCES

Chapter 1

page 16   (1)   Ward, Lester Frank. *Pure Sociology: A Treatise on the Origin and Spontaneous Development of Society.* New York: The Macmillan Company, 1903, pp. 171-184.

page 16   (2)   Fuller, Buckminster. *Ideas and Integrities.* New York: Prentice-Hall, Inc., 1963, p. 64. Since this was written, Fuller's new book *Synergetics: Explorations in the Geometry of Thinking,* New York: The Macmillan Company, 1975, has appeared. Fuller is the great pioneer in this field and deserves to be considered the founder of synergetics as a science.

page 17   (3)   Benedict, Ruth, quoted in A.H. Maslow's *The Farther Reaches of Human Nature.* New York: The Viking Press, 1971, p. 202.

page 17   (4)   Maslow, *op. cit.,* chapter 14.

page 17   (5)   Craig, James H., and Marguerite Craig. *Synergic Power: Beyond Domination and Permissiveness.* Berkeley, CA: Proactive Press, 1974.

page 17   (6)   Hampden-Turner, Charles. *From Poverty to Dignity: A Strategy for Poor Americans.* Grove City, NY: Anchor Press/Doubleday, 1974.

Chapter 2

page 28   (1)   Korzybski, A.L. *Science and Sanity,* 4th edition. Lakeville, CT: The International Non-Aristotelian Publishing Company, 1958.

Chapter 3

page 30   (1)   Skolimowski, Henryk. "The Twilight of Physical Descriptions and the Ascent of Normative Models" in *The World System,* Ervin Laszlo, ed. New York: George Braziller, Inc., copyright 1973 by Ervin Laszlo, pp. 97-119.

page 31   (2)   Merton, Robert K. *Social Theory and Social Structure.* Glencoe, IL: Free Press, 1957.

page 31   (3)   Theodorson, G.A., and A.G. Theodorson. *A Modern Dictionary of Sociology.* New York: Thomas Y. Crowell Company, 1969.

Chapter 5

page 46   (1)   Bellman, Richard. "Mathematical Models of the Mind," *Mathematical Biosciences* 1:287, 1967.

page 46   (2)   Platt, John R. *The Step to Man.* New York: John Wiley & Sons, Inc., copyright 1966, p. 175. Reprinted by permission of John Wiley & Sons, Inc.

page 49   (3)   The Association for Humanistic Psychology, 584 Page Street, San Francisco, CA 94117, has published a list of "growth centers" dedicated to the development of the human potential.

page 52   (4)   Theodorson and Theodorson, *op. cit.*

page 55   (5)   Arbib, M.A. *The Metaphorical Brain.* New York: John Wiley & Sons, Inc., 1972.

page 55   (6)   Fischer, R. "On Separateness and Ownness: An I-Self Dialogue," *Confinia Psychiatry* 15:165, 1972.

Chapter 6

page 63   (1)   Maslow, *op cit.,* pp. 43, 45.

page 63   (2)   Tart, Charles T. *Altered States of Consciousness.* New York: John Wiley & Sons, Inc., copyright 1969, pp. 1-2. Reprinted by permission of John Wiley & Sons, Inc.

Chapter 7

page 78   (1)   Ward, Henshaw. "Thobbery" from *Builders of Delusion.* New York: Bobbs-Merrill Co., 1931.

Chapter 8

page 92 (1) Platt, *op. cit.*, p. 151.
page 98 (2) Van Vogt, A.E. *The Players of Null-A.* New York: Berkeley Publishing Corp., 1974, p. 40.
page 99 (3) Korzybski, *op. cit.*
page 101 (4) Winckel, Fritz. *Music, Sound and Sensation: A Modern Exposition.* New York: Dover Publications, 1967.

Chapter 9

page 106 (1) Van Vogt, *op. cit.*, p. 171.
page 112 (2) Bergson, Henri. *Creative Evolution.* Authorized translation by Arthur Mitchell, Ph.D. New York: Henry Holt and Company, 1911.
page 114 (3) Bateson, G., D.D. Jackson, J. Haley, and J. Weakland. "Toward a Theory of Schizophrenia" *Behavioral Science* 1:251, 1956.

Chapter 11

page 133 (1) Fenichel, O. *The Psychoanalytic Theory of Neurosis.* New York: W.W. Norton & Co., 1945, p. 31.

Chapter 16

page 166 (1) Forster, E.M. "The Machine Stops" from *The Eternal Moment and Other Stories.* New York: Harcourt Brace Jovanovich, Inc., 1929.
page 167 (2) See, for example, Mumford, Lewis. *The Myth of the Machine: The Pentagon of Power.* New York: Harcourt Brace Jovanovich, Inc., 1970.

Chapter 17

pages 179, 180, 181, 184, 185 (1) Craig and Craig, *op. cit.*
pages 179, 185, 186, 187, 188, 189, 190 (2) Hampden-Turner, *op. cit.*

Chapter 17 (cont'd)

page 179   (3)   Theodorson and Theodorson, *op. cit.*
page 183   (4)   The Committee for the Future, Inc., 2325 Porter
                        Street, N.W., Washington, D.C. 20008.
page 185   (5)   See also: Hampden-Turner, Charles. *Radical
                        Man: The Process of Psycho-Social Development.*
                        New York: Doubleday-Anchor, 1971.

Chapter 18

page 191   (1)   Reich, Charles A. *The Greening of America.*
                        Copyright 1970 by Random House, New York.

INDEX

263